THERMOECONOMICS

A Thermodynamic Approach to Economics

John Bryant

Third Edition

© **VOCAT International Ltd**   **2012**

Published by

VOCAT International Ltd
10 Falconers Field
Harpenden
Herts
AL5 3ES
UK

Third Edition

ISBN 978-0-9562975-3-2

Front Cover: Getty Images.

Contents

Preface

This book, first published in 2009, stems from research that I began more than three decades ago when I was then working as group economist for the Babcock International Group. Prior to that, my formal university education had included degrees in engineering and management science – the latter in particular covering economics and operations research. What started out as a train of curiosity into parallels between the disciplines of economics and thermodynamics soon developed into something deeper.

Following publication of two peer-reviewed papers of mine on the subject in the journal *Energy Economics*, I was greatly encouraged in my research by other trans-disciplinary researchers with a similar interest, in particular, Dr László Kapolyi, who was then Minister for Industry of the Hungarian government, a member of the Hungarian Academy of Science and a member of the Club of Rome.

Not being based at a university and with no research grant at my disposal, my main thrust at that time had been to make a career as director of a consultancy and expert witness business and therefore, until more recently, opportunities to spend time on research had been few. Nevertheless, by the turn of the millennium I was able to find time alongside my consultancy to return to some research, and in 2007 published another peer-reviewed paper in the *International Journal of Exergy* entitled 'A Thermodynamic Theory of Economics', which was followed up with several working papers on monetary aspects and energy models. Interest in this work has been high, spurred on no doubt by general worldwide interest in energy and climate change.

This book and third edition is an attempt to bring together all the facets of the research into a coherent whole. Topics covered include the gas laws, the distribution of income, the 1st and 2nd Laws of Thermodynamics applied to economics, economic processes and elasticity, entropy and utility, production and consumption processes, reaction kinetics, empirical monetary analysis of the UK and USA economies, interest rates, discounted cash flow, bond yield and spread, unemployment, principles of entropy maximization and economic development, the cycle, empirical analysis of the relationship between world energy resources, climate change and economic output, and last aspects of sustainability.

Further developments have been added since the first and second editions, in particular, thoughts on production and entropy maximisation, order and disorder and relationships to the living world, which has necessitated re-organisation of some of the chapters. The chapter on money has been updated to incorporate empirical analyses of the recent upheavals in world economic activity from 2008 to 2011, though the conclusions reached have not changed, indeed, they have been reinforced.

The findings, interpretations and conclusions of this book are entirely those of my own, based on the research that I have conducted. While I have made every effort to be diligent and accurate, I cannot be responsible for the use that readers may make of my work, and they should satisfy themselves as to logic and veracity of the conclusions drawn.

I hope that this third edition represents an improvement and advancement on earlier editions, but would welcome nevertheless any feedback, discussions and corrections on points that readers may have.

I am indebted to my wife Alison for all her support and for providing an atmosphere conducive to my research.

John Bryant

CHAPTER 1   INTRODUCTION

The seeds of this book were sown back in 1974. At that time the Organisation of Petroleum Exporting Countries had cut oil production and raised the price of crude by a factor of four, from $3 to $12 per barrel, significantly affecting the cost of energy. There followed a period of virulent world inflation and then recession, but eventually economic life returned to 'normal' and growth resumed its apparent inexorable path. The early 1980s saw a recurrent bout of oil and price inflation and recession, with oil at one time approaching $40 per barrel. By 2008, two and a half decades later, oil prices had risen further, this time to over $140 per barrel, approaching fifty times the level of 1974 but, following the onset of severe recession, prices again fell back, though by 2011 had begun to climb again.

The history of oil and other fossil markets shows that the world economy has become evermore reliant on energy resources to maintain and fuel the economic machine and perhaps also growth in human population. This has prompted thoughts on the ways of working of economies with respect to energy, and over the last three decades the author began research into the links between energy and economics, with the thought that energy per se is not just a resource input, but that perhaps economic processes themselves reflect the thermodynamic laws governing energy systems.

The nature of the subject of this book requires significant proof for economists and scientists to accept that similarities between thermodynamic and economic phenomena might imply more than just a passing analogy or isomorphism. Indeed relations between the two disciplines have rarely been comfortable, with scientists sometimes having scant regard for the work of economists; and many economists believing that science has little to offer a discipline which, by its nature, can be thought of as anthropocentric rather than eco-centric. In his seminal book *'The Entropy Law and the Economic Process'* Nicholas Georgescu-Roegen (1971) opines that the science of economics is openly and constantly criticised by its own servants as being mechanistic. Thus even within disciplines there can be disagreement and opposing points of view. Trans-disciplinary research of this kind therefore is a hard path to tread. Despite these problems, however, similarities between thermodynamic and economic phenomena have caught the attention of a significant, growing band of economists and scientists.

2

1.1 Historical Research

Economist Paul Samuelson, in his book *Foundations of Economic Analysis* (1947), used the analogy of the Le Chatelier Principle and classical thermodynamics to explain the constrained maximisation problem, and acknowledged in his Nobel lecture (1970) that the relationships between pressure and volume in a thermodynamic system bear a striking similarity in terms of differentials to price and volume in an economic system. Other early contributors included Lisman (1949), who saw a similarity between money utility and entropy, Pikler (1954), who highlighted the connections between temperature and the velocity of circulation of money, and Soddy (1934), who suggested that if Marx had substituted the word energy for labour he might have conceived an energy theory rather than a labour theory of value. Perhaps the earliest suggestion of a relationship between the two disciplines was by Irving Fisher (1892), who related marginal utility to force, utility to energy and disutility to work. Fisher himself was mentored by Willard Gibbs, the founder of the theory of chemical thermodynamics.

The 1970s saw a rise in interest in the connections between economics and energy/thermodynamics, famously pioneered by Georgescu-Roegen in his book *'The Entropy Law and the Economic Process'* (1971). In a later contribution (1979) he noted that economic systems exchange both energy and matter with their environment and are best represented as open thermodynamic systems. He argued that the entropy law was important. Another strand of work, begun by Odum (1971, 1973), was to try to define the content of a product in energy terms, and he developed the term *"Embodied Energy"* as totalling the energy input into a product. This has met with some resistance within the economic community, because of the variability of the value and utility of money. Hannon (1973) also attempted to define an energy standard of value. The 1970s were also notable for the work of Meadows et al (1972) in *'Limits to Growth'*, a book which, while not concerned with relationships between thermodynamics and economics, certainly highlighted the potential links between economic output, resource consumption and pollution.

By the 1980s the position of energy as a contributor to economic output had risen significantly, and Ayres (1984) had begun his work on the impact of energy and exergy consumption on economic output. Costanza (1980) and Hannon (1989) were concerned with the 'mixed units' problem, commensurating dissimilar components. Costanza also argued for an embodied energy theory of economic value, similar to Odum. Other researchers at the time included Bryant (1982) on a thermodynamic approach to economics, Proops (1985) on general analogies between thermodynamics and economics, and Grubbeström (1985) on exergy. The decade finished with the work of Mirowski (1989) in his book *'More Heat than Light'*. His was a sustained attack on the foundations of modern neoclassical economics, that economics had copied the reigning physical theorems of the 1870s, with utility being a vector-field corresponding to energy.

In the 1990s, concern about the rising use of resources fired the work of Daly (1991, 1992), who posited the ideas of steady state economics and the relevance of entropy to the economics of natural resources. Söllner (1997), however, suggested that there was no direct link between thermodynamic properties and the characteristics of economic systems, and that there had been a failure of most attempts to produce economically interesting results.

The turn of the millennium saw a renewed interest in the connections between thermodynamics, energy and economics, spurred on no doubt by the advent of potential climate change and peak oil/gas. Candeal et al (2001) highlight similarities of utility to entropy, as do Smith and Foley (2002, 2004). These conclusions have been reinforced by Sousa and Domingos (2005, 2006), who also cite the Le Chatelier Principle in their work. They conclude that while neo-classical economics is based on the formulation of classical mechanics, economics is actually analogous to equilibrium thermodynamics. Chen (2002, 2007) has developed a thermodynamic theory of ecological economics from a biophysical point of view, encompassing non-equilibrium thermodynamics. Martinás (2002, 2005 & 2007) has also explored the idea of non-equilibrium economics, and stresses that microeconomics and thermodynamics are both based on the idea of exchange, but although irreversibility is a key part of thermodynamic analysis, this is not the case with neoclassical economics. Baumgärtner (2002, 2004) believes that the standard irreversibility concept of production theory is too weak to be in accordance with the laws of nature.

A number of researchers have concentrated their efforts in the area of econophysics, in particular, thermodynamic formulations of income and wealth distributions. Among these are Dragulescu & Yakovenko (2001), Ferrero (2004), Purica (2004), Yuqing (2006), Chakraborti & Patriarca (2008) and Chakrabarti & Chatterjee.

Ayres & Warr (2002, 2007) have produced several papers confirming the importance of exergy *(available energy or maximum useful work)* as a determinant of economic output, backed up by empirical research going back 100 years on the USA and UK economies. Ruth (2007) concludes that both conceptual analogies and attempts to quantify material and energy use from a thermodynamic perspective contribute nicely to the ongoing sustainability debate. Bryant (2007, 2008) has researched a thermodynamic theory of economics, with further papers, including empirical research, on monetary economics, peak oil and climate change.

From all of the above it can be seen that there exists a significant body of opinion that acknowledges that analogies or isomorphic links between the disciplines of thermodynamics and economics can be observed. Moreover the pace is quickening. This book and third edition is therefore an attempt to pull together all the above into a cohesive whole that can be recognised by both economists and scientists. In an effort to make trans-disciplinary research of this kind understandable and acceptable to people in both disciplines, a simplistic approach is sometimes required. The author therefore requests readers to bear with him if, on occasion, he appears to be preaching to the converted in one discipline or the other, it is not intentional.

Accepting the Darwinian principle of evolution, it is reasonable conclude that the human race is a product of the environment and the biological systems from which it evolved, and the ways in which it develops, including economic interaction, are therefore likely to reflect in some manner the ways in which nature and energy systems have evolved and operate. At the simplest level of the latter, the fundamental principle guiding the kinetics of reactions between chemical substances is the Le Chatelier Principle which states: *"If a change occurs in one of the factors under which a system is equilibrium, then the system will tend to adjust itself so as to annul as far as possible the effects of that change"*. Such reactions obey the laws of thermodynamics, in terms of heat production/consumption and the change in entropy arising.

At a higher level of entity, living organisms are composed of complex chemical compounds that interact with one another, but made up chiefly of molecules of oxygen, hydrogen, nitrogen and carbon, all of which can exist as gases (the last with hydrogen or oxygen). Goldberg et al (1993-1999) have collated thermodynamic data on enzyme-catalysed reactions.

Moving still further upwards, Schneider (1987) has pointed to Schrödinger's *'order from disorder'* premise (1944), which was an attempt to link biology with the theorems of thermodynamics, whereby a living organism maintains itself stationary at a fairly high level of orderliness (low level of entropy) by continually sucking orderliness from its environment. Schneider and Kay (1992, 1995) state that life can be viewed as a far-from-equilibrium, dissipative structure, that maintains its local level of organisation at the expense of producing entropy in the environment. Successful species are those that funnel energy into their own production and reproduction and contribute to autocatalytic processes thereby increasing the total dissipation of the ecosystem. Swenson (2000) has proposed a law of maximum entropy production stating that a system will select the path or assemblage of paths out of available paths that minimises the potential or maximises entropy production at the fastest rate given the constraints. This idea has been taken further, first by Mahulikar & Herwig (2004), with regard to the entropy principle applied to the creation and destruction of order, and second by Annila & Salthe (2009), regarding the idea of economic activity being an evolutionary process governed by the second law of thermodynamics.

While acknowledging the difficulties concerning the construction of analogues, it is, nevertheless, not a far-flung idea to propose that economic principles may have connections with and reflect the workings of natural phenomena and the laws of thermodynamics which govern all life, albeit at first sight economics and thermodynamics appear to be very different animals. Consequently it is of interest to examine the analogy and the extent or not to which it may pass the level of an isomorphism.

However, it is not enough just to accept that there may be connections. A thermodynamic representation of economic systems has to reflect the reality of how they operate in practice; the complex interconnecting systems of stock and flow processes carrying economic value from resources through to production, consumption and waste; and the feedback mechanisms of births and deaths, investment and depreciation, to replenish parts of the system such as population and capital stock. Such a representation must also have regard for the relationship of economics to resource availability, the environment and ecological systems. One might venture, vice-versa, that it is essential also that economics should pass muster with science, and relate properly to the actuality of natural and thermodynamic systems as they are and the way they operate.

1.2   Economics and the Ideal Gas

To connect the world of economics to that of thermodynamics, we turn to the analysis of gas systems. In the physical world, gases can absorb energy from a heat source with a higher temperature level, or by being compressed, raising their internal energy, resulting in a rise in temperature.

It might be argued at this point, that while even a small volume of a gas contains a very large number of molecules, homogenous and at first glance fairly evenly but chaotically dispersed *(Avogadro's or Loschmidt's number indicates 6 x 10^{23} in just a thimbleful)*, some economic systems by contrast can be composed of just a few different items, and unevenly dispersed. Clearly relationships derived from a theory applied to a small system might be significantly clouded by the problems of small-sample statistics. But the counter arguments are that many economic systems and markets are quite large and the problems of small sample statistics would not then apply. Moreover, economics take advantage of a human invention called money, a convenient commodity/medium of exchange with the property of linking non-homogenous economic factors together, so that they effectively work in a homogenous fashion.

To examine the characteristics of gases in more detail, recourse is made to the kinetic theory of an ideal gas. Some might argue that real gases are imperfect, and that their properties can diverge significantly from a model of an ideal gas. However, scientists take account of this fact by modifying the formulae arising from the notion of an ideal gas *(such as the 'compressibility factor' and Van del Waal's equation)* to enable thermodynamic principles to be applied more accurately. Economists also accomplish just the same in their own field, by developing econometric and statistical models encompassing a number of factors to explain the variations in the real world that they see.

The kinetic theory of gases teaches us that, for a closed ideal gas system made up of a number **N** of molecules, which are perfectly elastic and are busy moving about colliding with each other exchanging kinetic energy, the relationship of the system with the outside world is that it is contained in a volume **V** resulting in the gas exerting a pressure **P** on the walls of the system. If, through the application of heat from outside, the gas molecules are made to vibrate and move about faster, they increase their rate of exchange of kinetic energy and the gas accumulates internal energy resulting in a temperature rise **T,** with pressure and volume potentially increasing too; rather like a sealed balloon being heated and inflated. The relationship between the factors is given by the ideal gas equation:

$$PV = NkT \qquad (1.1)$$

Where **k** is called the Boltzmann Constant (Ludwig Boltzmann 1844-1906). Temperature **T** constitutes a measure of the relative kinetic energy level of the gas; the higher it is, the higher the velocities of the gas molecules, and the shorter the time between collisions of the gas particles with the walls of the system. Physicists utilise the concept of temperature by constructing a *scale* with reference to observable characteristics of physical things, such as the freezing and boiling points of water, the expansion and contraction of fluids and solids, and other phenomena. It provides a base to measure and venture further.

In thermodynamics, distinctions are made between flow and non-flow systems – see figure 1.1. For a non-flow system, such as a balloon or a piston cylinder, generally the number of gas units **N** is held constant, with pressure **P** and volume **V** being a function of temperature **T**. For flow systems, such as a pipe or a gas turbine, **N** becomes a flow of units per period of time (N_t), with a corresponding flow of gas volume V_t per unit of time; though varying with pressure **P** and temperature **T**. In both flow and non-flow gas systems the Boltzmann Constant **k** remains fixed. A distinction is also made between open and closed systems. In a closed system it is only possible for energy in the form of heat to cross the system boundary; matter itself cannot cross the boundary. In an open system, however, both heat and matter can cross the system boundary.

Non-Flow System   *Flow System*

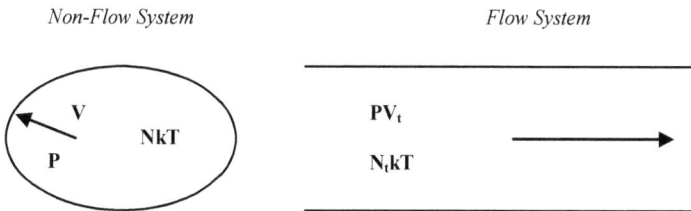

Figure 1.1 Thermodynamic Systems

In thermodynamic flow systems it is common to divide both sides of equation (1.1) by N_t, as in equation (1.2):

8

$$Pv = kT \qquad (1.2)$$

Where $v = V_t/N_t$, is the specific volume per molecule; the inverse of the gas density *(The more usual thermodynamic presentation of this format is to work in terms of the volume per mass of a very large number of molecules, but we will not confuse the issue here)*. This arrangement simplifies the kinetics of the analysis of a flow system, as the variable v becomes independent of time, because volume V_t and units N_t are both effectively flow measures per unit of time, with time therefore cancelling out.

Turning now to an economic stock system, an equation with a similar structure to equation (1.1) can be constructed. Imagine a system involving a number N of *'carriers or holders of value'*, where each carrier or holder can carry or hold a constant amount of embodied value or productive content k, *not* dependent on price or volume.

Clearly a concept such as embodied value or productive content might appeal to a scientist who is accustomed to measuring variables against absolute reference points. An economist might argue, however, and with some justification, that economics is not an absolute discipline, but a comparative one where value, or at least exchange of value, is not ascertained by deterministic processes used by scientists. It should be strongly emphasised, therefore, that by positing a productive content k, we are *not* implying that a scale of monetary 'productive content' can be constructed for a currency by reference to some independent, fixed reference level of value. Economics is very much a comparative discipline, and the value of one currency can and does change compared to another, arising from inflation and international comparisons. But we *are* stating that any non-monetary good or service is made up from a particular mix of non-monetary components that have a very specific productive content, however defined, which are immutable. A particular bolt has mass, is made of steel, which involved a type of energy transfer, and a long line of sub-sources of productive content. The fact that its price may change by virtue of substitutes or of demand, does not change the shape and content of the bolt in any way. Likewise one could define the content of a unit of currency as being $1, £1 or other notional value based on the confidence of its users, but this does not mean that this will forever have the same equivalence to the productive content of non-monetary goods; however that is measured for each.

A further point to state here with respect to the *nominal* value k, is that we are not ascribing a utility value, but a productive content, that is a *physical* value that a unit of economic stock possesses. A bolt still looks like a bolt, and a £1 note still looks like a £1 note. Utility, however, is a notion invented by economists to explain the paradox of say diamonds having much higher and potentially variable *prices* attached to them, through an exchange or trade, than can be explained in terms of the cost of their production or their usefulness, compared to say water. The process by which the notion of utility arises in a thermodynamic context will be described at a later chapter, though it should be noted that it has to do with entropy.

With the above in mind, for the time being we will put aside the problem of what standard the constant **k** for a particular product is to be measured against, i.e. energy, material content, labour man-hours or any other entity that might be regarded as a constituting a scale of reference. It is enough for the moment to assume that the result is acceptable to the parties in an economic system; otherwise they would not willingly trade with each other. However, a key difference to note between gas and economic systems is that whereas the constant **k** is fixed for the former, in the latter **k** differs from product to product. For example, even adding a few pages to a book changes its value of **k** (though its price may not change). Economic systems are composed of multitudes of products (and classes thereof), all with different value of **k**. Even money can be defined in terms of different currencies.

Returning to our discourse, the relationship of the system with the outside world is that the value held by the carriers or holders of value can be exchanged for goods and services, or the value held by other *different carriers or holders*, at the boundary of the system at price **P** and volume flow V_t over a period of time, and vice-versa, according to an *Index (or a degree of a scale) of Trading Value* T_t with which they can do this over that period. If they could increase their index of trading value T_t over the period, then the number of times the carrying units are re-cycled and used again could go up *and/or* the unit value of exchange of goods (the price) could also increase over the period. Thus the relationship of the variables is given by the ideal economic equation:

$$PV_t = NkT_t \qquad (1.3)$$

The above equation corresponds to the acknowledgement by Samuelson concerning the similarities between economic price and volume and thermodynamic pressure and volume, and to Pikler's remarks highlighting the connections between the velocity of circulation and temperature.

The index of trading value T_t has similarities with and is related to turnover, cost and added value, though the distinction is that while turnover, cost and added value can be defined in terms of a scale of value, rising or falling with respect to our index of trading value T_t, they are not technically the same as T_t, unless they are divided through by Nk.

While the index T_t is most readily equated to the velocity of circulation of a currency, there is no reason why it should not be compared also with the velocity of circulation of other items of exchange, such as the turnover of a producer stock, or the depreciation of capital stock. Even a labour force can be regarded as a stock, with new entrants coming from births through education, and retirals at the end of a working life. It is just that the lifetime comparisons are very different, from almost instantaneous for electronic money, to forty or fifty years for a member of the labour force; and much longer for some resources, if not noticeably depleted, and for some waste stocks, if not recycled back into the eco-system. In addition to velocity, however, the index T_t also carries a connection to the value of exchange (price P) compared to the productive content k.

Thus, by way of example, a producer may have a stock of identical finished items, which leave at a volume rate of V_t per unit of time at price P, with an equal and opposite flow of money from a customer to purchase the stock flow. The money flow represents a turnover of the money stock used to finance the operation, and likewise the value flow leaving the producer stock represents a turnover or velocity of circulation of the producer stock. The 3-dimensional plot of price P versus volume flow V_t and index of trading value T_t at figure 1.2 indicates that for a given index of trading value T_t the carriers could carry more or less products with lower or higher prices, and a change in the index of trading value T_t can give rise to a change in volume flow V_t, a change in price P or both.

It is important to stress that the index of trading value T_t so described here is one based on value, and not volume. If value flow, equal to price P multiplied by volume flow V_t, can vary on one side of the equation then on the other side of the equation value must be able to vary as well. Of the factors on the other side, the embodied value/productive content k that can be carried or held by a carrier, although inherently a value, is a *nominal value* and is deemed to be constant for that *particular* carrier. It is the same whether trading occurs or not. A £ of currency is still a £ of currency. A grain of wheat is still a grain of wheat, whether or not it is traded. Likewise, a barrel of a particular type of oil has weight, energy content and other properties which might be regarded as constant. As it is possible that the number **N** of carriers of value in a particular system configuration may be fixed (e.g. shares in issue), then the index of trading value T_t must be able to embody both changes in volume *and price* in order to make both sides of the equation compatible with one another.

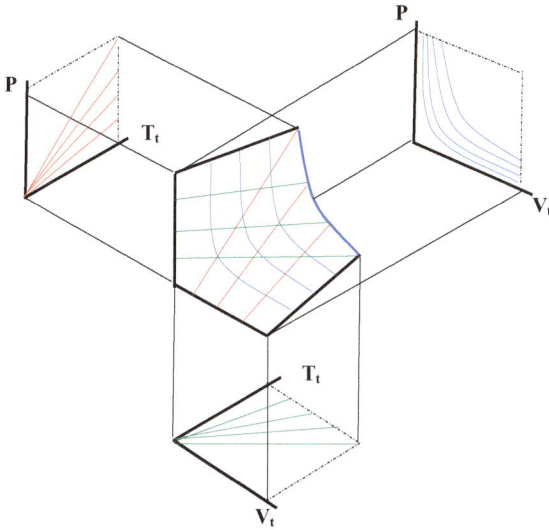

Figure 1.2 Price, volume flow and the index of trading value

The structure of the ideal economic equation can be clarified further by reference to dimensional analysis. In a thermodynamic system, at the boundary, pressure **P** is measured by force **F** per unit of area (length x length = L^2) on which it acts (i.e. **F** x L^{-2}), and energy **J** is a product of force x distance moved (i.e. **F** x **L**). Thus pressure **P** is equivalent to energy **J** per unit of volume (i.e. **J** x L^{-3}). The Boltzmann constant **k** is defined as energy (**J**) per molecule per degree of temperature (**T**). Therefore restating equation (1.1) in simplified dimensional terms we have:

$$\left(\frac{J}{L^3}\right) \times \left(L^3\right) = N \times \left(\frac{J}{NT}\right) \times T \qquad (1.4)$$

Similarly for an economic system, at the boundary, in dimensional terms, price **P** is measured as total value flow **J** per volume flow V_t of items (i.e. value/unit of flow), and the embodied value or productive content **k** is measured as value **J** per carrier per index (or degree) of trading value T_t. Hence re-stating equation (1.3) we have a similar dimensional presentation:

12

$$\left(\frac{J}{V_t}\right) \times (V_t) = N \times \left(\frac{J}{NT_t}\right) \times T_t \qquad (1.5)$$

There are two main differences between to the two systems.

First, the gas system is defined by the volume containing the energy of the gas, L^3 in dimensional terms. It is spatial and 3-dimensional. In an economic system, however, volume flow V_t does not have a dimensional configuration, and the value contained by the economic unit can be said to act at a 'point' with no spatial dimensional format. It may be a pin, a bank note or a power station, but it is still considered to be acting at a 'point'. This aspect does not matter, however, as the economic value J is likewise defined as per item 'point' flow, and not per spatial volume as in a gas system, and therefore L^3 and L^{-3} at equation (1.4) effectively cancel each other out for the economic system. The first difference of the two systems is illustrated at figure 1.3.

Gas – 3-Dimensional Volume   Economic – Point

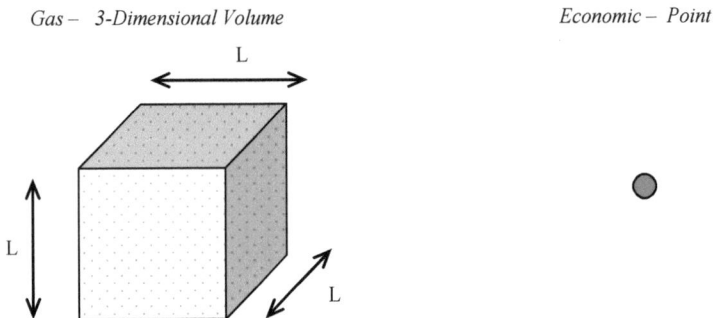

Figure 1.3 Gas and Economic Formats

The second difference is that of throughput flow and time. In a gas *flow* system, throughput flow is defined by reference to volume flow V_t per unit of time on the left-hand side of the equation, and flow of molecules N_t per unit of time on the other side; as shown in figure 1.1. The same time dimension occurs on both sides of the equation and is of the same measure, that is V_t and N_t proceed in tandem together. In a non-flow gas system, on the other hand, time does not enter into consideration, and the number N of molecules remains the same; and though the volume V can change through expansion and compression, it is not flowing in the sense of continually changing in content. Thus the ratios $V_t/N_t = v$ and $V/N = v$ retain the same relationship to each other via the specific volume and density, which relates to the volume L^3 format as in equation (1.2) **Pv=kT**.

Economic systems, however, have elements of *both* flow and non-flow processes. Thus on the left hand side of the equation we might envisage a volume flow throughput V_t, retaining the same time relationship (items per transaction time – a year etc) as that of a thermodynamic flow process, such as inputs and outputs from a stock. But on the other side of the equation the stock quantity **N** is ordinarily not flowing. It can of course change in size, according to the difference between input and output flows, but otherwise it stays where it is. On the right hand side of the equation therefore, the notion of flow is transferred to the index of trading value T_t, which becomes a velocity of circulation relative to the central stock **N**, and is related to *both* the *lifetime* of a stock item and the *price* of exchange. By re-arranging equation (1.3) we have:

$$P\left(\frac{V_t}{N}\right) = kT_t \qquad (1.6)$$

$$Pv_t = kT_t \qquad (1.7)$$

$$T_t = \left(\frac{Pv_t}{k}\right) \qquad (1.8)$$

Where the volume throughput rate per unit of stock per unit of time $v_t=V_t/N$ is inversely related to the lifetime t_L of an economic item in the stock. It will be noted that a change in the index of trading value T_t can be occasioned by a change in price **P** *and/or* a change in the volume throughput rate per unit of stock per unit of time v_t.

Figure 1.4 illustrates the lifetime principle. Two economic items are presented with different lifetimes: t_{L1} and t_{L2}. The first for example might represent money, having a shorter lifetime than a reference transaction period of time t_t of a year. Thus money gets turned over perhaps several times in a year. The second entity might represent some form of capital stock, with a lifetime of several years. Thus the volume flow in a transaction year will represent only a proportion of the lifetime of the economic entity, the depreciation or consumption rate *[A thermodynamic engine, by contrast, generally has a very small value of t_L compared to t_t].*

$$t_{L1}$$
$$\longleftrightarrow$$

$$t_t$$
$$\longleftarrow\!\!\longrightarrow$$

$$t_{L2}$$
$$\longleftarrow\!\!\!\longrightarrow$$

Figure 1.4 Lifetime of an economic good compared to the transaction time

Thus instead of a specific volume **v** with dimensions of $\mathbf{L}^3$, as posited for a gas system (equation (1.2)), an economic system has a volume throughput rate $\mathbf{v}_t$ per unit of stock per unit of time, with no spatial dimensions, but with a time dimension (equations (1.7) and (1.8)). To this is added the effect of price **P** relative to the productive content **k**.

The description of the variable $\mathbf{v}_t$ 'volume throughput rate per unit of stock per unit of time' is rather long, and for the rest of this book we will refer to it as the *Specific Volume Rate* **v**, being equal to volume flow **V** divided by stock quantity **N**. We will also drop the subscript t from the variables volume flow **V**, specific volume rate **v**, and index of trading value **T**, as these are automatically associated with a *flow* per unit of time.

From all of the above analysis it can be seen that the formats of the ideal gas equation and the ideal economic equation outlined so far are similar, with a defined equivalence; pressure **P** with price per unit, volume **V** with units of output/consumption per unit of time, the number of molecules of gas **N** with the number of particular carriers or holders of value in a stock, temperature **T** with the index of trading value, and the Boltzmann constant **k** with the embodied value/productive content per unit of the particular carrier and the index *(degree)* of trading value. In both gas and economic systems time is balanced out on both sides of the equation. The analogy suggests that value in an economic system might, in a manner to be determined, have some relation to heat content in a thermodynamic system. Figure 1.5 further illustrates the principle of economic systems.

The definition of open or closed systems changes with how the boundary is conceived. An economist might argue that a producer trading with customers and suppliers is an open system, but an economy that does not trade with any other economy might be regarded as a closed system. Following this argument, the world economy could be described as a closed system, as net trading would be zero. A scientist, on the other hand, would argue that all economic systems derive their benefit from the productive content found in the ground, sea and air, from the living things that grow and inhabit the earth, and from the energy supplied free from the sun. From a scientific point of view, no economic system can be regarded as being closed.

Irrespective of being open or closed, the real size of an economic system (net of inflation) is determined by the flow of volume and real value per unit of time.

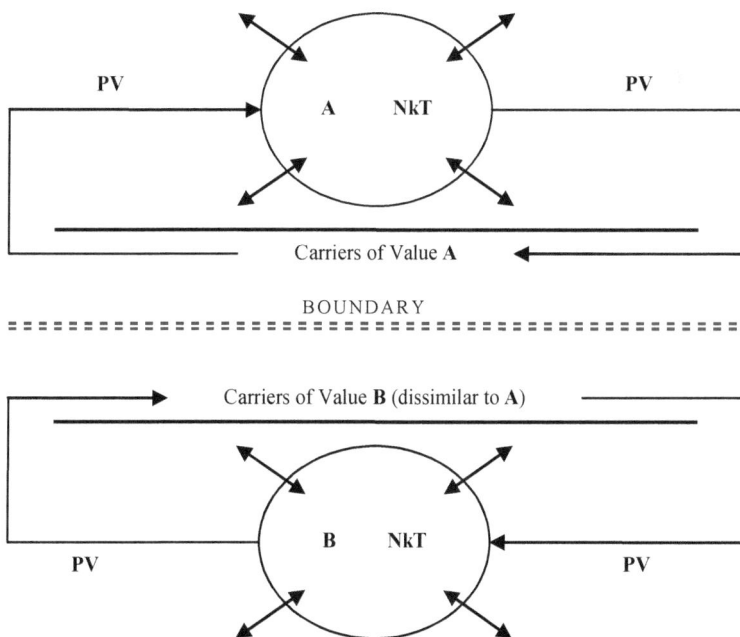

Figure 1.5 Economic System

It should be emphasised that if two sets of carriers of value are different in nature (e.g. money versus product output) then neither can cross the system boundary and they flow in the opposite direction to each other, though, as with thermodynamic systems, value can be exchanged between them. If, however, the carriers of value and the input/outputs are *one and the same* (such as a flow of products into a finished stock and then outputted to customers), then there is no boundary and they all flow in the same direction and are part of the same stock.

We now turn to consideration of the structure of economic stock and flow processes.

CHAPTER 2   STOCK AND FLOW PROCESSES

Economic systems are made up of multitudes of stocks connected by flow processes; starting with resources and ending with population, capital and consumer stocks and waste. While they are all different, there are nevertheless some characteristics that they share. Stocks have inputs and outputs, they have lifetimes of varying lengths, and through interconnecting flows they can interact with one another. There are also some characteristic stock families to which most can be compared to. Chief among these are money, human stocks, resource stocks, and producer and consumer capital stocks. Consideration should also be given as to how each stock interacts with the natural cycle.

The key economic flows connecting stocks are those involving production and consumption. Production brings together streams from resource, labour and capital stocks to produce economic output, and consumption (so called) brings together streams of economic output to nurture human life and reproduction and create consumer stocks. Figure 2.1 is a representation of a simplified economic system.

It may be seen that the envelope of the money stock has been drawn around familiar economic entities, but excludes the natural cycle. Not all aspects of resource stocks are considered in an economic system. For example, the ecological impact of reducing fish reserves on the well-being of the ocean eco-system is not traditionally considered to be a part of economic analysis, but the lack of fish stocks, impacting on supply and demand and on other sources of food pertinent to humans, is considered to be part of economic analysis. Likewise, until recent years, the impact of greenhouses gases on the atmosphere has not been considered by economists as being part of economic analysis. Opinion on these matters is however changing and links between economic activity and natural cycles are beginning to be recognised.

We now turn to examine the structure of a general stock and flow model, from which particular families will be derived.

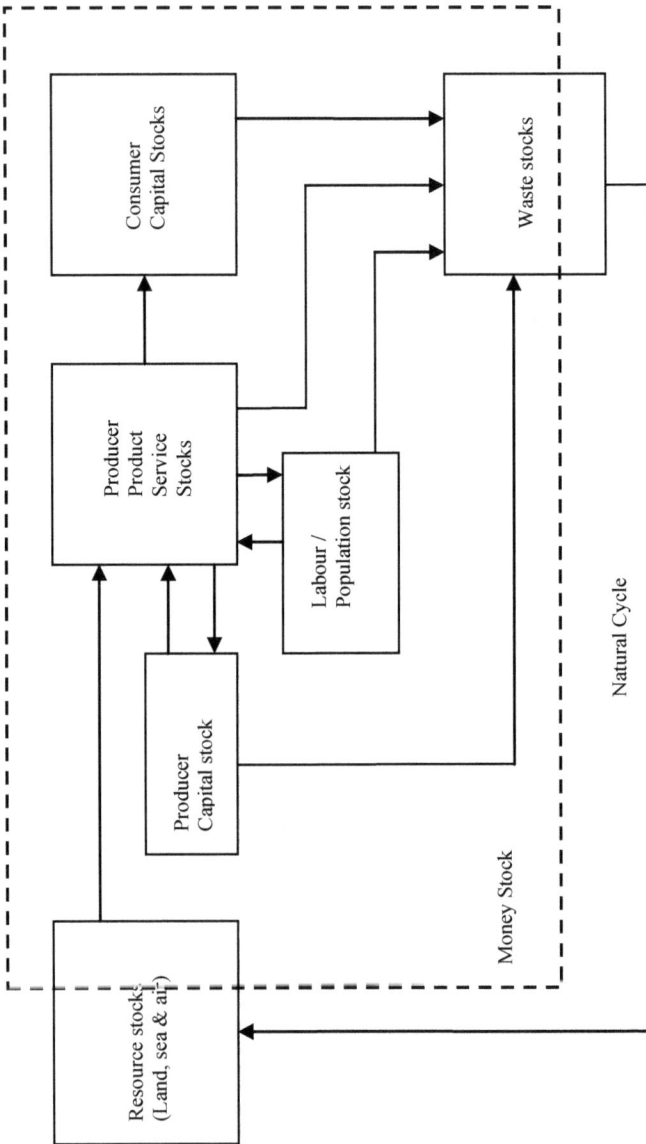

Figure 2.1 Simplified Factor Conversion Process

2.1   General Stock and Flow Model

A model can be constructed that follows the format of equation (1.3), which describes the process for a general stock. Imagine a stock composed of N_Z units of a good Z, each having a *nominal* value or productive content k_Z per unit, independently measured and not dependent upon supply/demand. Each unit of stock has a lifetime t_{LZ}. The stock of N_Z units has a *volume* throughput rate of V_Z per unit of time, equal to n_Z units of stock per transaction time t_t. The transaction time t_t is commonly regarded as a year or other time denomination.

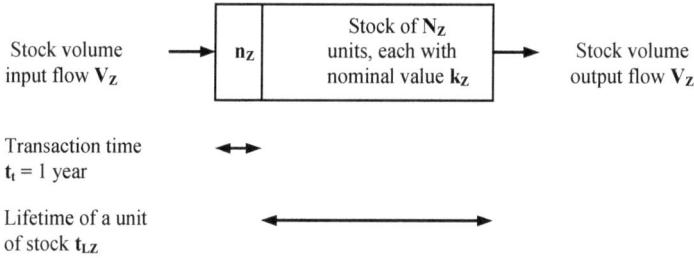

Figure 2.2 A Simple Stock Model

Thus a constant volume throughput V_Z per unit of time will be equal to:

$$V_Z = \frac{n_Z}{t_t} = \frac{N_Z}{t_{LZ}} = v_Z N_Z \qquad (2.1)$$

Where v_Z is the specific volume rate (the volume throughput rate per unit of stock per unit of time). Further, if we substitute also ξ as equal to the ratio of the lifetime t_{LZ} of a unit of stock compared to the standard transaction time t_t, and likewise, $\phi = (1/\xi)$ as the frequency ratio of the stock, we can also write:

$$\frac{t_{LZ}}{t_t} = \frac{N_Z}{n_Z} = \xi = \frac{1}{\phi} \; ; \qquad t_{LZ} = \xi t_t \; ; \quad t_t = \phi t_{LZ} \; ;$$

$$N_Z = \xi n_Z \; ; \quad n_Z = \phi N_Z \;\; \text{and} \;\; v_Z = \left(\frac{1}{\xi t_t} \right) = \left(\frac{\phi}{t_t} \right)$$

$$(2.2)$$

For a standard transaction time t_t of 1 year, ξ would therefore be a multiple of years, or a fraction of a year, and may be variable depending upon the flow rate. It is dimensionless, being only a ratio. Likewise the frequency ratio ϕ, being the inverse of the ratio ξ, is also dimensionless, measuring the relative frequency of a stock cycle compared to that for the standard transaction time t_t. Referring to the diagram at figure (2.2), it can be seen that if there is a difference between the volume flows into and out of the stock, say V_{Zin} and V_{Zout}, then there will also be relative changes to the dimensionless factors ξ and ϕ, when measured from each end of the stock.

There are arguments as to the natures of the lifetime ratio ξ and the frequency ratio ϕ of a stock. Some stocks maintain a constant productive value all the way through their lifetime, until consumed. Humankind however does not generally achieve productive potential before reaching adulthood, and the time of retiring from the labour force is variable and depends upon a number of factors. Though capital stocks may be designed for a particular lifetime, demand can dictate that they be scrapped early or have their lifetimes extended. Repair and maintenance and health expenditure tend to extend lifetimes of humankind and capital stock.

Now in our model each unit of stock has a nominal embodied value or productive content k_Z measured in terms of what it can do, or pass on when it is consumed. For example, fuels and minerals will have an inherent energy content (technically exergy with reference to the environment value) [Ayres & Warr (2003)], and food sources such as animate stocks, crops and agricultural land, will have vitamin, nutrient and energy content. These values are independent of any price that an economic system may attach to each unit of stock at particular times, though they can be improved by refinement, as in ore mining or petroleum plants, to remove non-relevant materials; but in such cases they effectively become another stock with a different value of k_Z. These values are *independent* of any exchange value or price P that may be attached to them by human producers and consumers.

It is more difficult to imagine the concept of productive content of human labour. First, the potential productive content of humankind is not constant, generally growing in the early years through the inputs of education and experience, before declining with old age. Second, particularly in developed countries, humankind has progressed from being primarily a workhorse, inputting energy to produce output, to being a factor supplying mostly information and management/ownership to the process. Even construction and farm workers often now use machines driven by energy sources in their work to produce output (rather than animal and human power), and their role now involves inputting information via hand controls. This assumes of course that a source of energy will always be there in the future to fuel this work. Third, some humans tend to work harder than others. Fourth, some humans generate more value than others do. Fifth, some humans do no work at all but receive an income from a fund of money or value, by virtue of accepted ownership or deferred wage (pension). Irrespective of these points, however, economics values labour power by what it is paid in money terms, which is reflected in the price or wage rate of labour. The latter valuation thus becomes a subjective assessment by humans as to the contribution they make to the production process and what they can reasonably acquire to their benefit.

Turning to inanimate resource stocks, from a thermodynamic perspective the embodied value or productive content to be abstracted from renewable or non-renewable resources technically exists already – such as the heat content of fuels, the metal content of ores, sunlight and wind (the latter two flows of energy delivered by natural forces). The productive content of energy resources is called exergy, and measures their relative value distance from the environmental average.

Evaluating the productive content of animate stocks, such as arable crops, domestic animals and fish, is a more complex exercise, though in principle it is no different to that applied to resource stocks. It is not the cost of the stock that delivers benefits to humankind and the economic process, but the inherent productive content such as nutrient and energy value. In addition, food production involves other inputs such as energy consumption in transportation, farming and packaging of food.

We must be careful to distinguish between energy and exergy. When energy is 'consumed' in a process such as in a power station, according to the law of conservation of energy, it is not destroyed, but converted from a high grade, high temperature source to a lower grade, lower temperature sink, with useful work such as electricity being abstracted. There is no loss of energy value. There is however an irrevocable loss of exergy value and an associated increase in entropy released to the environment, resulting from efficiency losses. Similarly a high grade iron ore has a greater exergy value than a low grade iron ore, and oil from a gusher oil well might have more exergy value than tarsand, because low grade resources require consumption of significantly more energy input to make them useable in the market. The refinement of ores involves the consumption of energy, to remove impurities and non-relevant materials to arrive at a refined but smaller more concentrated and useable amount of ore.

Much research has been carried out on the subject of exergy [Wall (1986); Warr, Schandl and Ayres (2007)], highlighting the net useful work that is obtained from the consumption of exergy in a process. Warr et al indicate that the aggregate fuel exergy conversion to useful work in the UK has risen from about 5% in 1900 to about 15% in 2000, which still leaves an 85% loss. They also point out that the exergy content of human muscle power is a very small component compared to that of other sources of energy.

An important point to make here is that in economic analysis, exergy losses and associated entropy gains are mostly not included in the GDP account (excepting that through economic entities operating specific waste recycling processes). These losses are substantial. It is assumed that nature absorbs them, as indeed it provided the resources in the first place.

Proceeding further, while net exergy content or value net of a waste function may represent the inherent productive content that can be abstracted from a resource relative to the value distance from the environmental average, an economic agent instead calculates the *costs* and mark-up it thinks it can recover by the way of a price. Economic value is set out in a two-sided structure, with expenditure on goods and services (the price) on one side, and income such as wages and profits on the other, with a net import/export adjustment to take account of inter-country trading. These items are defined in monetary terms and not resource productive content terms. Thus from an *economic perspective* ultimately resource consumption is defined in terms of money value of labour and capital stock; that is the cost flowing in the opposite direction to the stock flow, being an imputed 'ownership' of the stock flow, split into a wage share and a profit share.

To follow an example: imports of oil can be regarded as resource consumption expenditure, having productive content measured in tonnes and heat value per tonne. The oil has a price (per tonne or barrel) attached to it representing the (variable) money exchange value of the productive content. The economic money value of expenditure of the oil is represented on the other side in terms of the bought in costs, transport costs, wage costs and capital stock costs (profit) and any particular market conditions involved to produce and transport the oil. Bought in costs and transport also involve labour and capital stock costs and energy costs, and energy supply used up in transport in turn involves labour and capital stock costs, and capital stock production also involves labour costs - and so *ad infinitum.* These costs travel in the opposite direction to the productive content. A consumer does not comprehend the myriads of productive contents that went into the litres of petrol that he/she bought, only the cost per litre made up of wage costs and profit attributed or imputed to humans (capital stock is ultimately owned by humans).

Another example is that of consumption of capital stocks. Capital stock arises from new investment, the productive content of which is made up of material and information content, designed by human ingenuity and put together in the production process by humans and machines driven by energy. Clearly a definition of the productive content of one item of capital stock, let alone the millions of different forms it takes, from a power station to a pin, is complex. As with other parts of the economic process, economics get around the complexity by subjectively defining the capital stock on one side in terms of the expenditure of money paid for new investment, whatever form or size it takes and its productive content, and on the other side by the money value the owners of the capital stock are deemed to possess, be it a house, a factory or a piece of land. But, when looking at the source from whence the new investment in capital stock came, this is represented in terms of the wages paid to the humans who constructed the capital stock investment, dividends and rent to the human owners of the production unit, and the rest from payments for external resources and consumption of capital equipment, which in turn, taking the process one step further back, come from other processes also incorporating wage costs; even the owner of an oil field charges a wage or dividend for his/her trouble for the energy transported to power the capital equipment. Thus, in the ultimate, all capital stock models are effectively calculated on the basis largely of accumulated human wages, pensions, dividends and rent from ownership, which travel in the opposite direction to the productive content added. This is not to deny that real value in terms of productive content and exergy is abstracted from the capital stock as it is used up. It is just that economics approaches the valuation of the consumption of productive content in a different manner; through the assumed costs which travel in the opposite direction to the productive content, but ignoring mostly the exergy and environmental losses that nature absorbs.

Ultimately, resources are economically free to humankind – nature does not sell them to us. Economics makes the assumption that humankind has effectively commandeered resources, and is distributing them to itself, according to a perceived self-interest benefit and the labour and capital stock cost of acquiring/producing them. While such a process may not be in principle different from say a bird building a nest, commandeering twigs, moss and mud in the process, the scale involved is of course much greater. Thus there is a fundamental difference between economic and

24

thermodynamic perspectives. Figures 2.3, 2.4, 2.5 and 2.6 set out representations of the process, and the reconciliation of productive content with economic output.

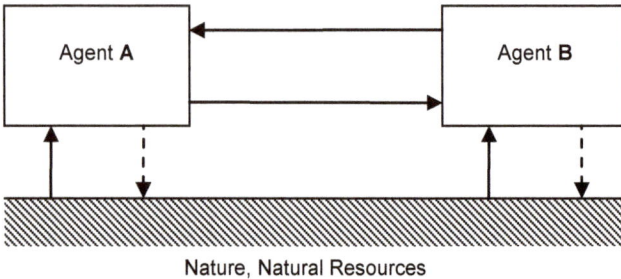

Figure 2.3 Economic Distribution and Exchange of Resources

At figure 2.3 economic agents **A** and **B** abstract productive content from resources, which are free, bar the work done and means used to abstract them, and exchange economic value between each other so that each may share, in some agreed manner, of the fruits of the earth. At figure 2.4, as primary productive content is inputted to production processes, at each stage waste of non-relevant productive content is generated, which mostly is not reflected in economic accounting presentations: for example, surface mining of coal, creating spoil heaps, followed by production of electricity, with waste energy going up the chimney.

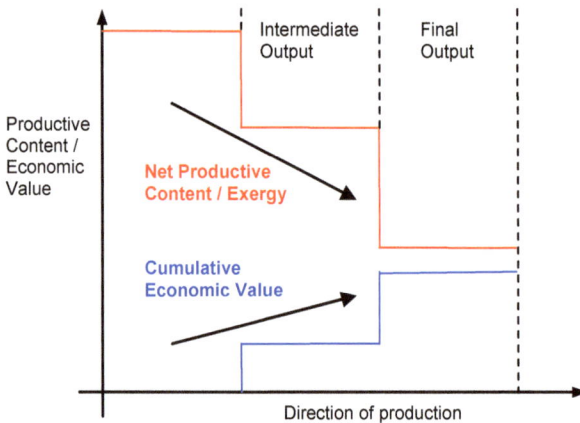

Figure 2.4 Economic Accumulation of Value, & Refinement of Productive Content

Figure 2.5 reconciles the relative differences between economic and productive content, and figure 2.6 illustrates the shape of an economic system.

The chart at figure 2.5 indicates that the contribution of human work is small compared to the total of productive content consumed, yet human wage accounts for much of economic output. This arises because on an economic basis humankind assumes 'ownership' of resources, without having to 'recompense' resources for any losses of value.

Now the author would among the first to admit that the economic approach of evaluating everything in terms of a common value denominator called money is a wonderful thing, which conveniently simplifies the problems of comparison and barter, and makes for a very efficient means of running an economic system. One should not, however, lose sight of the productive content of units of economic stock, and its impact and origins with respect to the eco-system, as opposed to the varying monetary values that may be attached to it.

Even so, it is the case that exchange values *are* attached to outputs at all stages of the economic production process (net of exergy losses), as they proceed through the economic system towards final output and demand, and it might be supposed that such values reflect to a degree the productive content accumulated inside; or at least those values humankind is prepared to associate with them, and less the wastage and efficiency losses at each stage. Thus a matrix model in the manner of an input-output table [Leontief (1986)] would imply productive content within it as well as economic value. It is reasonable therefore to suppose that while economic flows in and out of a particular *stock* will do so in accordance with the thermodynamic laws and have the same productive content **k** per unit going in and coming out; in an economic *production/consumption* process between successive stocks, however, thermodynamic losses are ignored, though at each stage *net* productive content **k** per unit is added to/refined (as in a production process) or reduced (as in a consumption process).

26

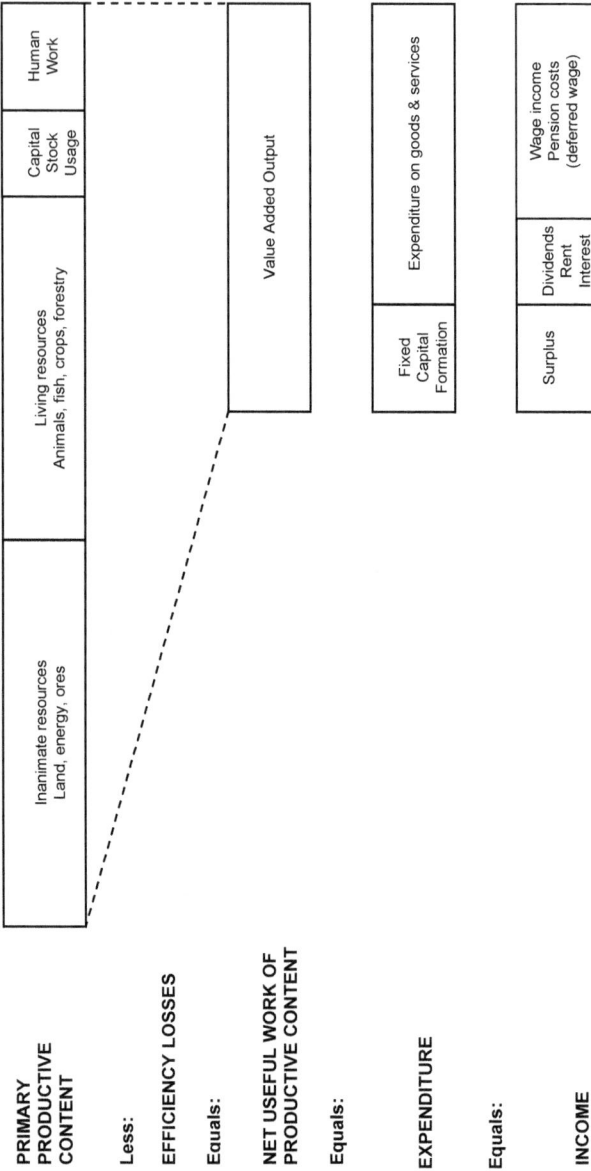

*Figure 2.5 Simplified Reconciliation of Productive
Content with Economic Output*

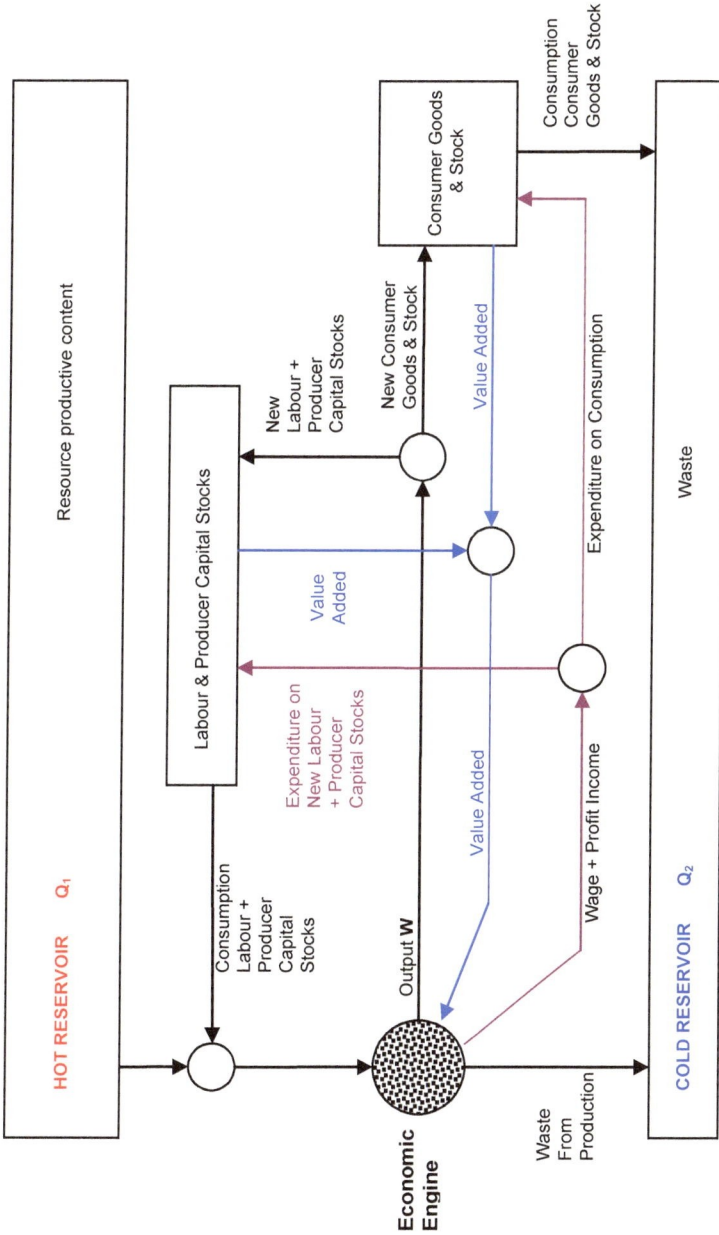

Figure 2.6 A Simple Economic Cycle

Now we might imagine that the value of economic output flow rate G_M in *monetary* terms will be equated to its price multiplied by a volume rate of flow V_Z as per equation (2.1):

$$G_M = P_M V_Z \qquad (2.3)$$

The volume flow rate here equates to the final output exergy value sold on to customers at each stage, as depicted at figure 2.4; for example, barrels of oil sold by an oil field owner to an oil company, tonnes of coal sold by a mining company delivered to a power station, and units of electricity sold to a distributor. At each stage the sales value is calculated by reference to a price multiplied by the final economic volume supplied, and the added value by reference to sales less economic input costs. Mostly, however, losses through waste and irreversible processes are excluded.

However, alongside our economic monetary output value function G_M, which emphasises the human contribution, there is another function representing the output value flow rate in terms of thermodynamic productive content, which we will call G_O, and travelling in the opposite direction to money.

To reconcile economic concepts with thermo-economic ones we imagine that at a particular point in time in production the two output value flow rates are deemed to be equivalent to each other, though of course they are defined in very different ways:

$$G_M \equiv G_O \qquad (2.4)$$

G_O is defined in terms of some measure or complex of measures of final productive content. It is immutable. That is how it was made/processed (ready for sale), harvested or dug out of the ground. G_M on the other hand is defined in terms of human money costs and the relative price, which can vary according to supply and demand (though further complicated because variable delays can occur between receipt of productive content and payment for it in terms of money). Thus, at every stage along the economic production chain between stocks, a loss of value (as per the Second Law of thermodynamics) occurs in terms of productive content, yet no loss in money output value has been assumed. Therefore, at the macro level, one can imagine an economy to have a Second Law production efficiency ratio, according to the amount of productive content lost, in the manner of figure 2.5.

Suppose we let G_T represent the input productive content value flow prior to all the production processes, and G_X the lost value, then we may write at a particular point in time:

$$G_T - G_X = G_O \equiv G_M \qquad (2.5)$$

The above is an equivalence relationship, as different units of measurement are involved; £ versus KCals of energy, $ versus nutrient value of grains of rice, or similar.

We could further set out the contributions of each of the components of output (resources, labour input and consumption of capital stock) but, whereas in terms of money we could specify the costs involved in an additive manner (resource cost, wage cost and depreciation), we cannot readily do this in terms of productive content and the efficiency losses incurred, primarily because productive content is defined in different ways, not easily reconcilable – protein value against metal content or calorific value, for example. In general terms therefore we might write:

$$\left[G_O = f\left(G_T - G_X \right)_{R,L,K} \right] \equiv \left[G_M = \left(G_R + G_L + G_K \right) \right] \quad (2.6)$$

which sets out a difference between the productive contents of inputs and the irreversible losses incurred in a production process, equating to the net productive value / exergy of the output G_O, which in turn is equivalent to the money value of final output G_M proceeding in the opposite direction (bar any possible time differences). In the alternative we could write:

$$\left[G_O = f\left(\eta G_T \right)_{R,L,K} \right] \equiv \left[G_M = \left(G_R + G_L + G_K \right) \right] \qquad (2.7)$$

where η represents a composite production efficiency factor to convert input productive content to net output.

In setting out the above we have not posited a formal additive equivalence of output flow of productive content / exergy to economic value flow over time, but we could say that since the flow of productive content on the left hand side of the equation must obey the laws of nature, in particular the laws of thermodynamics, then the flow of economic value on the right hand must also comply with those laws.

It is widely acknowledged that a universal scale of productive content is unlikely to be a feasible proposition, however much a scientist might otherwise wish it. Some goods may be more amenable than others, but even energy, which might be thought of as the most likely candidate, carries with it a whole host of problems, such as the relative costs of international transport, local taxation, and the quality and energy value of each energy source and product. However, humankind, while acknowledging the many criteria associated with each good, including nutrient value, energy content, weight, size, functionality, lifetime, convenience etc, chooses to employ a utility/pricing process to connect them all together. Nevertheless, although a universal scale of productive content will likely never be achieved, it is constructive to recognise the differences, and to take account of this in assessing economic flows. Three important conclusions of the above analysis are as follows:

- While the productive content **k** per unit entering or leaving a particular, defined stock is the same, that accumulating through intervening production/consumption processes changes, with (mostly) undefined losses occurring at each process.

- In a given production process having output value flow rates defined in terms of both productive content flow rate G_O and economic value flow rate G_M (the left and right hand sides of equation (2.7)), it will not likely be easy or practical to determine the absolute and proportions of the efficiency losses incurred by consumption of each input component – resources, labour or capital plant though it may be possible to examine a specific component to calculate the exergy loss/entropy gain, such as fuel usage in a power station, but not to compare it to that of all the other inputs, such as labour and capital stock depreciation.

- Third, if the flow rates of both economic value and productive content are proceeding in tandem with each other, it is reasonable to posit that if changes in productive content / exergy losses occur such that the overall value flow rate of the left hand side of equation (2.7) goes up or down, then the overall economic value of the right hand side of the equation may also change, in sympathy with what is happening on the left hand side. Thus it may still be possible to consider the thermodynamic effects of overall *changes* to the flow rates of economic final output value for each good or service, and ultimately for an economy, even if the absolute productive content / exergy losses incurred during successive production/consumption processes cannot be calculated.

2.2   Monetary Stock Model

Returning to our development, and excluding for the moment a comparison with a monetary standard, equation (1.3) for a stock could be written as:

$$P_Z V_Z = N_Z k_Z T_Z \tag{2.8}$$

which expresses the price and volume flow of inputs and outputs in terms of the number of stock units N_Z, the nominal inherent productive content k_Z per unit, and the index of trading value T_Z. It will be noted here that everything on both sides of the equation is expressed in terms of the stock Z. Thus, in an ideal world with perfect economic comparison, price P_Z would be equivalent to net productive content k_Z, and the volume throughput V_Z per unit of time would be equivalent to stock units N_Z multiplied by the index of trading value T_Z per unit of time. It is of course quite possible that, when connected to other systems, the market might take a different view of the value of productive content k_Z, and price accordingly. Thus the market might view a box of 100 widgets with 100 widgets of productive content (made up of the productive content from consumption of capital stock, labour and resources) as being worth more or less than this. Consequently both price P_Z and specific volume rate $v_Z = V_Z/N_Z$ can vary with respect to T_Z.

To compare the stock with other, different stocks, and to facilitate the trading process, economics introduces the concept of money, in the manner of the opposing flow system at figure 2.7.

Equation (2.8) for the economic value flow rate G_M then becomes:

$$G_M = P_M V_Z = N_M k_M T_M \tag{2.9}$$

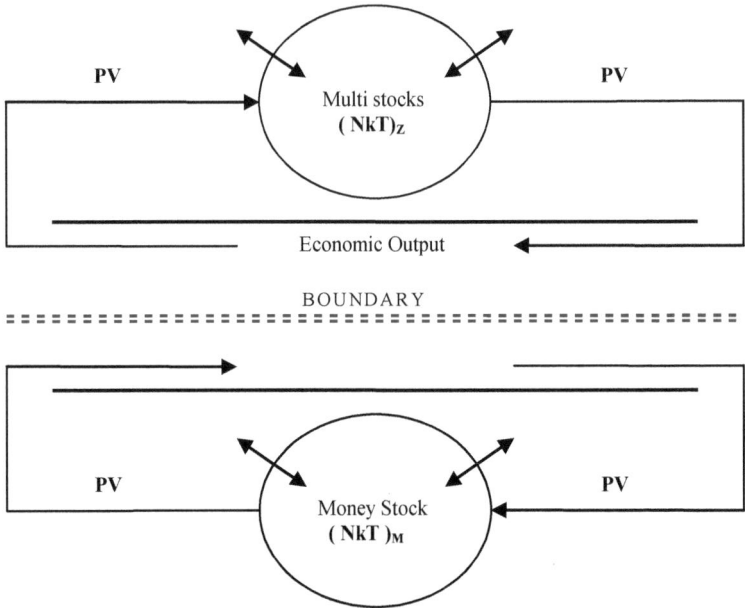

Figure 2.7 A Monetary System

Where all factors except volume throughput V_Z of stock Z are instead expressed in money terms, and where T_M is the index of trading value (velocity of circulation) of the financing money stock. It should be noted that the two stocks are only comparable through the interface of input and output, and that they flow in the *opposite direction* to each other. By combining equations (2.8) and (2.9) we have:

$$\left(\frac{P_Z}{P_M}\right) = \left(\frac{N_Z}{N_M}\right)\left(\frac{k_Z}{k_M}\right)\left(\frac{T_Z}{T_M}\right)$$

But since the embodied value or productive content of *both* currency k_M and stock units k_Z can be defined as 1 *[£1, 1 car (of a specific make and model) etc]*, then the equation reduces to:

$$\left(\frac{P_Z}{P_M}\right) = \left(\frac{N_Z}{N_M}\right)\left(\frac{T_Z}{T_M}\right) \qquad (2.10)$$

Which expresses the relative prices in terms of the relative stocks and indices of trading value (velocities of circulation), and noting that the flow from one stock is in the opposite direction to the flow from the other stock.

Thus for example, suppose stock Z of a small business was composed of two identical PC computers each valued at 1 computer unit of productive content (if such were a standard), with no inflation or utility/entropy gain. Thus price P_Z of output is equal to k_Z, which is equal to 1. The computers lose value at 50% per annum straight-line depreciation and are replaced every 2 years. Thus the right hand side of equation (2.8) becomes 2 x 1 x $0.5 = 1.0_Z$, which also equals the left hand side - the depreciation of the computers. Thus the small firm has to buy one new computer each year. We now choose to value computers in terms of currency M. We suppose that the price P_M of 1 computer unit is equal to 1,000 M units of currency. Thus the left hand side of equation (2.9) becomes $1,000_M$ x $1.0_Z = 1,000_M$. It is a matter of practicality as to what funding a small business might have, either in terms of positive cash reserves in a bank or of borrowed funds. If the small business kept reserves of 6 months trading, not unlike the money supply of an economy, then the amount of reserves earmarked for computer replacement would be given by:

$$1000_M = N_M \text{ x } 1_M \text{ x } 2$$

Thence the nominal stock value of currency $(N_M k_M)$ funding computer replacement would be 500_M, to which would be added other amounts to fund the rest of the business.

Hence a thermodynamic model of a monetary system, joining all the sectors, will have the form:

$$PV = NkT \tag{2.11}$$

Where P is the overall price index of an economy, V is output per annum at constant prices, N is the supply of money (MO, M2, M3, M4 definitions etc), k is a monetary constant (£1, €1 etc) and T is the money velocity of circulation. In this model, output of products and services from a stock *[the left-hand side of equation (2.11)]* flow in one direction, and money *[the right-hand side of the equation]* flows in the *opposite* direction.

It is intuitively obvious that a higher level of trading index (velocity of circulation) T is associated with a shorter-lifetime of money. Electronic money has a very short lifetime, whereas commercial paper and some bonds (included in M3) have maturity up to 5 years. Though it is traditional to consider money stock as being composed of money instruments up to 5 years' maturity, there is no reason why this should not be expanded to include long term debt, such as mortgages, corporate and government bonds, though the dynamics of such a system might be somewhat different to one with a narrower base of money stock.

Referring back to the gas model set out in the introduction to this book, a higher level of temperature is associated with a higher kinetic energy, which is a function of the average velocity of molecules, and hence inversely proportional to a function of the lifetime between collisions of molecules with the boundaries of a system. Thus the thermodynamic arrangement of an ideal gas still fits with the essential modus operandi of conventional monetary Quantity Theory and the perceived workings of a monetary system. Chapter 5 of this book is devoted to developing in more detail the thermodynamic concepts of a money system, including elasticity, entropy and interest rates.

It should be strongly emphasised again that by setting out the above analysis we are *not* implying that an *absolute* scale of monetary 'productive content' can be constructed for a currency, in the manner of exergy, or any other scale of independent value. Economics is very much a comparative discipline, and values of currency can and do change, arising from inflation and international comparisons. But we *are* stating that any non-monetary good in a stock has immutable productive content per unit, and that a currency is defined as having a *nominal* value $1, £1. It is thus a matter of confidence and monetary control as to whether the nominal value of money matches the productive content of the goods travelling in the opposite direction, or whether changes in price and money balances occur to offset any mismatch.

2.3   Labour Sector

Human labour activity constitutes a major input to all economic systems, though its nature has changed from historical past, when it provided essentially only 'man-power' (with the assistance of domesticated animal power). In the modern economy much power is now provided by machines which consume energy, some of which is non-renewable in the scale of human existence. Human labour has graduated somewhat therefore towards providing, utilising and managing information in support of production processes. Without a source of energy to fuel the machines, the modern economy would have difficulty in operating and the nature of human labour would almost certainly have to change in form.

As in the general stock model at section 2.1, as production of output progresses over time, a proportion of the labour force that provided the 'human-power' and 'information-power' is consumed (retired). This equates to the throughput volume flow rate **V**. New units of labour will eventually enter the system to offset this reduction, via the births/upbringing cycle of population, *provided* that sufficient fruits of output are directed towards the costs of births/upbringing. Demand will also influence factors such as recruitment levels, redundancies, unemployment and early retirement, which affect total labour requirements. It is of course quite reasonable to expand the model by reference to links with changes in population cycle of births, education and retirement, though that is set aside here.

In labour terms equations (2.1) and (2.8) for *one* stock can be combined and expressed as follows:

$$P_L V_L = P_L \left(v_L N_L \right) = N_L k_L T_L \qquad (2.12)$$

Where P_L is the effective unit price of labour throughput, V_L is the labour volume throughput, v_L is the specific volume rate (the throughput rate of labour per unit of labour stock per unit of time), N_L is the total number of labour units existing, k_L is the *nominal* value or productive content that a unit of labour can provide over a lifetime, and T_L is the index of trading/velocity of throughput of labour. Thus the left-hand side of the equation represents the labour cost consumed or generated per unit of time, and the right-hand side the effective turnover of the labour productive content per unit of time.

It is intuitively obvious that in a single stock process, if the price of labour P_L is fixed as equalling its productive content k_L, then the specific volume rate v_L is equal to the index of trading value T_L. It is of course quite possible that, when connected to other systems, the market might take a different view of the value of labour productive content k_L, and price accordingly. Consequently both price P_L and the index of trading value T_L can vary with respect to the specific volume rate $v_L = V_L / N_L$.

In economics it is common to write labour cost as wage rate w_L paid per unit of time multiplied by total labour units N_L in use, thus $w_L N_L$.

Substituting this into equation (2.12), then the left-hand and right hand sides become:

$$P_L V_L = P_L \left(v_L N_L \right) = w_L N_L = N_L k_L T_L \qquad (2.13)$$

Where w_L, the wage rate per unit of time, is equal to P_L, the wage cost per unit of labour throughput, multiplied by v_L, the specific volume rate of labour per unit of time. Eliminating we have:

$$P_L \left(v_L \right) = w_L = k_L T_L \qquad (2.14)$$

Hence the wage rate w_L is a function of the productive content k_L multiplied by the labour index of trading value T_L.

It will be noted that if output on the left-hand side of equation (2.13) goes down, this results in a reduction in $w_L N_L$. The net effect is therefore either a reduction in the wage rate w_L or a reduction in the employed labour force N_L, the latter implying a rise in the level of unemployment.

We now introduce a money stock alongside the labour stock to connect it to the other parts of an economic system. The flow of the money stock is in the opposite direction to the flow of the consumption of labour. As set out in the general stock model for our economic system and postulated at equation (2.9), there exists an equal and opposite flow of money to labour output (cost), and we can therefore write as an alternative:

$$w_M N_L = N_M k_M T_M \qquad (2.15)$$

Where the wage rate w_M per unit of time is instead measured in terms of money, N_M is the supply of money to the labour process, k_M is the nominal value of a money unit, and T_M the velocity of circulation of money to the labour sector. Thence, because k_M and k_L are nominally expressed as 1 (£1, productive content one person), we can write:

$$\left(\frac{w_L}{w_M} \right) = \left(\frac{N_L}{N_M} \right) \left(\frac{T_L}{T_M} \right) \qquad (2.16)$$

While it is the accepted norm to measure wages and prices by reference to units of money, this does not negate the principle of the underlying lifetime value or productive content of a unit of labour – however that may be measured.

The money stock with respect to labour is connected to all the other money stocks to form a central pool. From our development earlier in this chapter it was pointed out that because of the way economic accounting systems are constructed, the wage element does not just account for the input of human-power, it effectively accounts also for much of other inputs, such as resources *[see figures 2.3 - 2.6]*. National income accounts include, on one side, wage income, profit surplus and net imports/exports and, on the other side, expenditure in cost terms. They do not specify the productive content and exergy of resources. Where are the millions of tonnes of oil, steel and all the other inputs? Answer: in the wage cost. Thus the wage/income rate largely encompasses much of the total system production, not just the real productive input provided by human-power. We could thus express equation (2.14) as:

$$P_E v_E = w_E = k_M T_E \qquad (2.17)$$

Where T_E is an overall wage index of trading value, and reflects not only the throughput of human-power and relative prices, but the accumulated value from resources and other inputs. T_E is not just a velocity of circulation with respect to labour therefore, but is a measure of the overall value of output per capita of an economic system that humankind attributes to itself. Since k_M is automatically given the value 1 (£1, $1 etc), T_E is effectively represented by the wage rate w_E, which would be a *mean* or *average* rate.

It is well known that there is a distribution of wealth and earnings about a mean value. Figure 2.8 is an example of an earnings distribution, which exhibits a skewed curve, flattening and widening with an increase in the mean value.

From a historical perspective one common form of parametric distribution used by economists to measure the frequency distribution of income in an economic system is the Lognormal distribution [Klein (1962)], and the density function on a scale of wage rate w_i is given by equation (2.18), where μ is the mean value of log w_i, and σ is the standard deviation. The general shape of this distribution is that of a skewed curve.

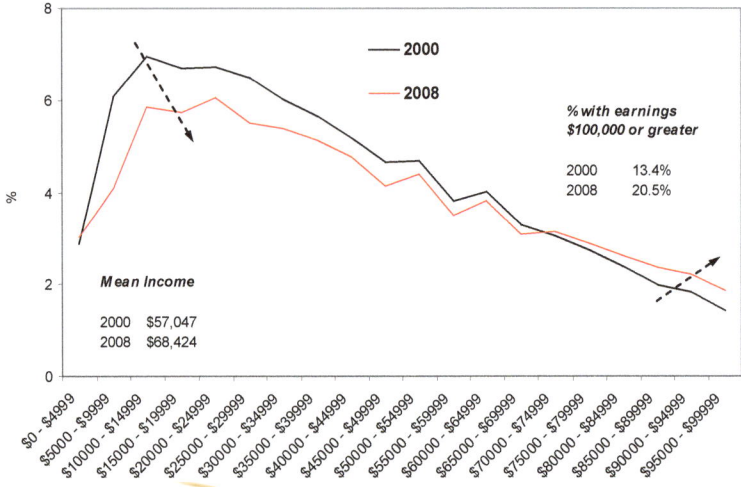

USA Census Bureau Current Population Survey: Tables HINC-06, HINC-07

Figure 2.8 Income Distribution USA 2000 and 2008

$$f(w_i) = \left(\frac{1}{2\pi}\right)\left(\frac{1}{w_i\sigma}\right)e^{-\left(\frac{(\log w_i - \mu)^2}{2\sigma^2}\right)} \tag{2.18}$$

However, a number of researchers of the Econophysics school of thought have highlighted also the possibility of using variants of the gamma distribution to measure the distribution of income and wealth. They include Yuqing (2006), Ferrero, Purica (2004), Chakraborti et al (2008), Dragulescu & Yakovenko (2000) and Chakrabarti & Chatterjee and others. Chakrabarti & Chatterjee have proposed an ideal gas-like market model, with per capita money being a function of temperature **T**, and Ferrero has proposed a similar model, closer to the Maxwell Boltzmann distribution.

From the analysis set out so far in this book, it is likely that a similar distribution will occur. Equation (2.19) sets out the frequency formula for the Maxwell Boltzmann Distribution as applied to an ideal gas system operating in 3 dimensions:

$$f(w_i) = \left(\frac{2}{\sqrt{\pi}}\right)\left(\frac{1}{kT}\right)\left(\sqrt{\frac{w_i}{kT}}\right)e^{-\left(\frac{w_i}{kT}\right)} \qquad (2.19)$$

Where w_i is the energy of a particle, k is the Boltzmann constant and T is temperature. Figure 2.9 sets out a chart of the equation for varying values of temperature T.

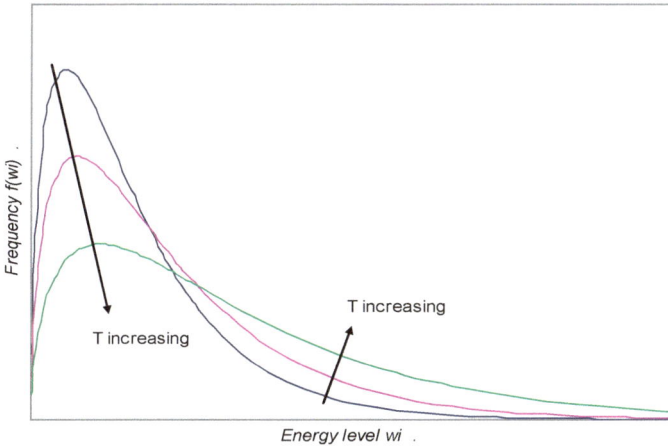

Figure 2.9 Maxwell Boltzmann Distribution for an Ideal Gas

We could imagine therefore that a distribution of a similar form might apply in an economic system, with w_i instead representing individual wage rate, and $w_E = k_M T_E$ representing the mean wage. Thus, since the productive content k_M is constant (£1, $1 etc), the mean wage w_E is equal to the index of trading value T_E (the economic temperature). It will be recalled however that in the opening chapter it was noted that an economic system does not have a three-dimensional format, and value was considered to act at a point. To allow for potential differences for this effect, while retaining the skewed nature of the distributions experienced in real life, a reasonable approach is to postulate a more general gamma distribution as in equation (2.20).

$$f(w_i) = A\left(\frac{(w_i)^\theta}{(kT)^\mu}\right)e^{-\left(\frac{w_i}{kT}\right)} \quad \text{or} \quad f(w_i) = A\left(\frac{(w_i)^\theta}{(w_E)^\mu}\right)e^{-\left(\frac{w_i}{w_E}\right)}$$

$$(2.20)$$

where w_E is the mean or average wage rate, and A, θ and μ are constants to be determined.

It will noted from figure 2.9 that as temperature T in a gas system rises, raising the speed and energy of gas molecules, there is a shift in the distribution of energy values, with those at the lower end declining and those at the higher end increasing. There is also a shift to the right as the average energy level kT increases.

If it is accepted that there is a significant similarity between economic and thermodynamic systems, then the following conclusions might be drawn. First, just as gas particles with high energy are more likely to collide more often with the walls of a system than those with low energy, in a similar reasoning, in our ideal economic system some individuals with higher activity rates and value impact on economic output (by reason of their attributes) may carry higher value than others. Thus wage or income rates, which measure the money value generated per unit of time, may vary in a similar way.

Second, from equation (2.17) it can be seen that the mean wage rate w_E is equivalent to $k_M T_E$, the productive content k multiplied by the index of trading value T. Thus, if income distribution followed a general gamma or Maxwell Boltzmann distribution, it might be expected that a rise in the index of trading value T would entail a rise in wage rates, and shift the curves to the right.

The chart at figure 2.8 shows a similar effect in terms of USA income distribution from 2000 to 2008, with the lower end declining and the higher end increasing; and with mean income shifting to the right.

2.4   Stocks of Economic Output

Stocks of economic output occur in many parts of the economic system. Producer capital stock serves to produce and enhance output by manipulating energy and materials to make things in a manner and speed that a human being alone could not match. Consumer capital stock, such as houses and transport, serves to assist and enhance the lifestyle of human beings. Items of consumer expenditure, from food to services and other items serve to sustain and enhance human life and lifestyle. Producer stocks represent potential output from the economic system. All are part of the same family, but are on the expenditure side of an economy, opposite to the income flow of wages.

There are two main measures that define stocks of economic output. The first is the relative average productive content arising from the sheer volume and variety of each of the products and services available, literally millions, which may vary from that of a few grains of rice to that of a house or a power station. This contrasts to the scale of humanity, which is composed of people (male and female) whose potential productive content must come within a much narrower band of variation. The second measure is that of the lifetime of an item of output; from a bottle of milk which might last only days, even if refrigerated, to a house, which might last more than a hundred years. One might contrast this with the scale of humanity, where life expectancies are in a much narrower band (say from 20 years to a century). Figure 2.10 sets out a chart of stock classification with some examples.

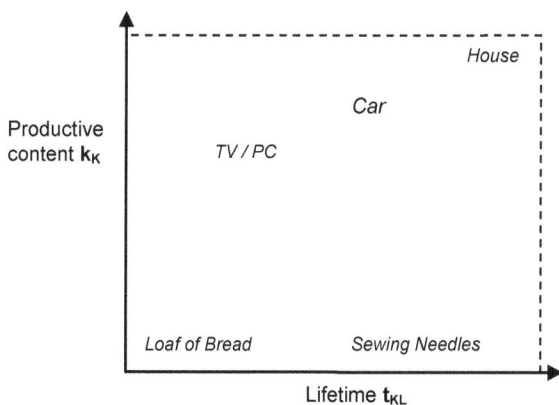

Figure 2.10 Economic Stock Classification

It might be imagined that capital stocks should be defined as a separate category of economic stock, because of the long lifetimes typically attached to them. This does not altogether accord however with human experience as even some very small items can have long lifetimes. Thus for example, while a large power station might be said to have a lifetime of say 50 years, the same might also be true of a packet of sewing needles residing in the home of a human. Being of small value, however, economics tends to write off the lifetime of sewing needles and define them to be part of consumption. Items are generally classed as capital stock when their embodied value/productive content is large *and* has a long lifetime,

42

whereby the value can be written off gradually over time in a set of accounts.

Equations describing economic stocks can be set out in a similar manner to the general stock model at section 2.1 of this chapter. We will use the subscript **K**, primarily to denote capital stock, though the derivation can be applied to any economic output stock, any differences being related to relative lifetime and productive content. In stock throughput terms equation (2.8) can be written as:

$$P_K V_K = P_K \left(v_K N_K \right) = N_K k_K T_K \qquad (2.21)$$

Where P_K is the effective price/cost/value of a unit of stock throughput, V_K is the volume throughput per unit of time of stock units (not value), N_K is the total number of stock units, v_K is the specific volume rate, k_K is the *nominal* embodied value/productive content of a unit of stock and T_K is the index of trading value of the stock. The specific volume rate v_K is most readily associated with the depreciation rate of a stock, and is inversely related to the lifetime of the stock. The left-hand side of the equation therefore represents flows to and from a stock, and the right-hand side nominal stock value $(N_K k_K)$ multiplied by an index of trading value T_K, which is a function of both price and volume flow. Differences between inputs and output rates to the stock will be associated with residual increases/decreases in total stock size.

In a similar fashion to the general stock model, a reverse monetary flow equation can be set out.

$$P_M V_K = N_M k_M T_M \qquad (2.22)$$

With the price of each economic item then being defined in terms of money units.

Here, however, there is a divergence, in that items that come within general consumption and of short lifetime could be set alongside a short term money definition, say M2 or M4, whereas items that have a long lifetime, such as a house financed by a mortgage or a capital investment by long-term finance, require to be set against a wider definition of money. Thus in a similar manner for the general stock, equation (2.10) becomes:

$$\left(\frac{P_K}{P_M}\right) = \left(\frac{N_K}{N_M}\right)\left(\frac{T_K}{T_M}\right) \tag{2.23}$$

Now by re-arranging equation (2.21) and eliminating stock quantity N, we can write:

$$P_K v_K = k_K T_K \tag{2.24}$$

Relating price P_K and specific volume rate v_K to the productive content k_K and the index of trading value T_K. The specific volume rate v_K is of course most readily associated with the consumption rate of a stock, and is inversely related to the lifetime t_{KL} of stock units. The format of this equation is similar to that for the wage rate (see equation (2.14) $w_L = kT$), but substituting $P_K v_K$ for w_L. Thus $P_K v_K$ will represent the value throughput rate per unit of stock, and is situated on the expenditure side of an economy, compared to the wage rate, which is on the income side. We will call $P_K v_K$ the *Value Rate*. Clearly, if we could summate all the throughputs of the many economic items flowing, this might give rise to a pattern of rates; perhaps related in some way to the pattern of wage rates coming in the opposite direction. We are however jumping the gun a little in our exposition. Figure 2.11 sets out a more detailed diagrammatic presentation of figure 2.10 for economic stocks. The top diagram relates price P_K to the specific volume rate v_K, and the bottom diagram sets out the relationship to lifetime t_{KL}, which is the inverse of v_K.

As an example one might classify some items shown at figure 2.10 as in table 2.1, using some very broad brush estimates of economic value and lifetime.

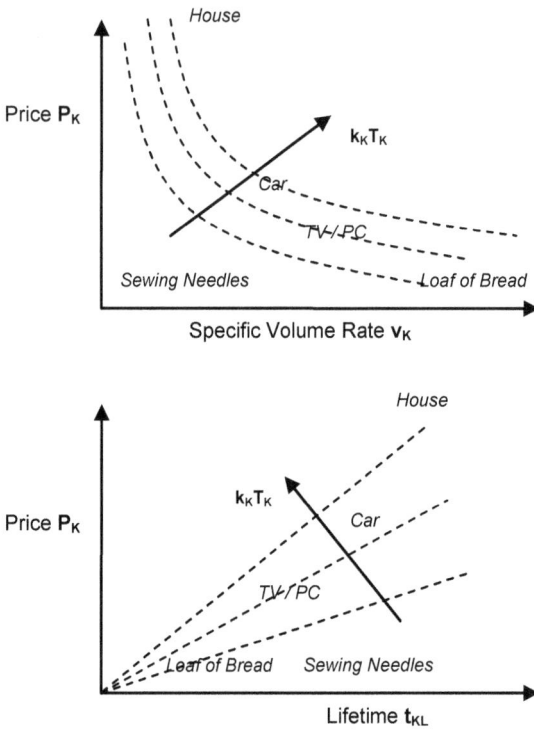

Figure 2.11 Value spread for items of economic stock

However, knowledge of average transaction prices and lifetimes is largely unknown, except at the micro-economic level. The lifetime of fixed investment, involving items such as dwellings, factories, roads, rail, transport equipment, machinery, must be long, perhaps 20 years or more. Durable goods, which include furniture, cars, telephone equipment, can last over a period of considerably more than a year. Semi-durable goods, which include clothing, recording media etc, also can last more than a year, but not as long as durable goods. Non-durable goods, such as food, drink and vehicle fuels, generally last less than a year, though a lifetime of a year might be imputed to seasonal food, such as bread and vegetables.

Table 2.1 A Scalar Approximation of Some Types of Economic Items.

	Lifetime t_L	Specific Volume Rate $v=1/t_L$	Capital Cost P	Value Flow Rate p.a. Pv (depreciation)
Large Power Station	50 yrs	0.0200	£1000m	£20m
House	80 yrs	0.0125	£200K	£2.5K
Car	12 yrs	0.0833	£15K	£1.25K
TV / PC	5 yrs	0.2000	£400	£80
Loaf of Bread +	1 yr	1.0000	£1	£1
Packet Sewing Needles	50 yrs	0.0200	£2	4p

+ *Seasonal – grain comes once a year*

Clearly all the items are very different in nature.

Another approach is to examine national accounts, which provide a limited classification of some expenditure items. The following items of annual expenditure relate to the UK for 2007.

	£bn
Gross domestic fixed capital formation	249.2
Durable goods	96.4
Semi-durable goods	93.8
Non-durable goods	209.8
Government Expenditure	296.3
Services	447.3
Total	**1,392.8**

UK Blue Book 2008 edition

Of particular note however is that around half of expenditure involves government and services, which do not have a purely physical form. The output is not *made* as in manufacture, farming or mining. One has only to think of items such as repair & cleaning, hospitals, postal services, defence, education and local and central government, to recognise this. As an example, a patient in hospital receives the care of doctors and nurses (wages), medicines (manufactured goods), hospital depreciation (capital stock), warmth (energy consumption) and many other inputs, but the output is (hopefully) a patient who has been made well again. In a service, a distribution of values will more likely relate more to the inputs to the service, rather than the output. And as the inputs are likely to be several in number, as in the example, their distribution will be spread out over a number of sectors. Thus a distribution of expenditure values in government and service functions will reflect the cost inputs on the other side of the accounts, rather than the output, notably of course wage rates. A concept of a distribution of expenditure value therefore has only limited meaning.

2.5   Resource Stocks

It might be supposed that a resource stock could be regarded as having only an output, any input material coming from land, sea or air. This is not technically correct however. Mining requires productive content to be used up to dig a hole in the ground, fishing requires consumption of productive content to send a boat out to sea to gather fish in a net, oil from a well requires a shaft to be sunk, and productive content to be used up in production and exploration, and water from a desalination plant requires consumption of energy and other productive content. All these activities, however, are production processes, to convert a resource into a useable product, involving a consumption of a variety of productive contents at different efficiency rates. There is, nevertheless, one feature that differentiates resource stocks from other economic stocks, which is that resource stocks may or may or not have a regenerative/replenishment function coming from nature or the sun. Thus an oil well will eventually run dry, and a fish stock can be depleted if the time and pace of the natural cycle is ignored. The oceans, however, might be regarded as infinite with respect to desalination processes. Figure 2.12 illustrates the principles.

Clearly if a resource stock does not have a replenishment function, notated by input volume V_{R1}, then, as output from the stock proceeds, the number of units N_{R1} of the stock declines until, if demand continues, there are no stock units left. For non-renewable stocks such as oil reserves, a replenishment function might be served by an exploration process, but in the ultimate at a world level the latter will not be enough if demand and production continues to expand. In between these types of stock are semi-renewable stocks, of which examples are fish stocks and agricultural land. As long as demand is offset by replenishment, then the process can continue. If agricultural land is over-farmed, this will result in reduced crop yields. Economic processes do not concern themselves with replenishment, except where human-made regulating processes exist to provide this, such as periodically allowing farm fields to lay fallow or international agreements to restrict sea-fishing.

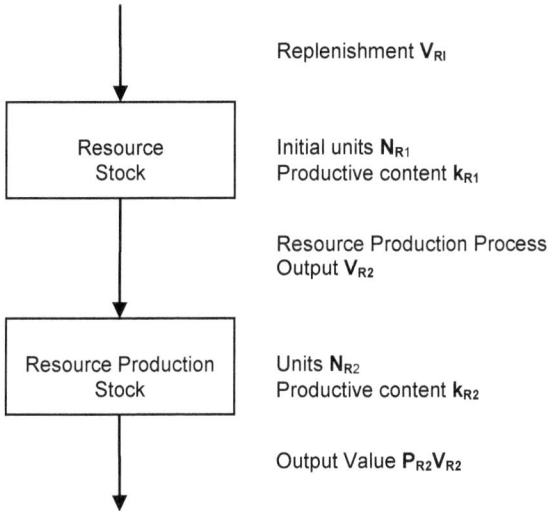

Figure 2.12 Resource Stock

For an output process **O** equation (2.8) can be stated as:

$$P_{RO}V_{RO} = P_{RO}\left(v_{RO}N_R\right) = N_R k_R T_{RO}$$

Where P_{RO} is the effective price/cost/value of a unit of resource stock output at stage 2 in figure 2.12, V_{RO} is the volume of resource stock unit throughput per unit of time, N_R is the total number of resource stock units, v_{RO} is the resource stock specific volume rate (a volume depletion, extraction rate), k_R is the *nominal* lifetime value or productive content of a unit of resource stock, and T_R is the index of trading value of a resource stock, which is a function of both price **P** and specific volume rate *v*. Factors of production will include productive content of workers, energy, and consumption of capital stock to mine/extract/harvest a resource, and productive content of fertilisers and mining technology/exploration to improve yields. In an identical manner as before we can express resource usage in terms of opposite flows of money associated with it:

$$P_M V_{RO} = P_M (v_R N_R) = N_M k_M T_M \qquad (2.25)$$

In economic terms the above money equation relates to resources proceeding to destinations outside the system, such as imports or purchases from another economic system. Thus money is received from purchasers and is used to fund the wages of resource workers, energy costs, depreciation costs, and profits of mine owners. As noted earlier in this chapter, while resources have productive content, measured in terms of exergy or other inherent value, which is passed on to purchasers, the money returning in payment goes to fund human labour and capital stock money values attached to them. It clearly does not pass to the original resource.

2.6   Environment Waste Stocks

Waste stocks come in three broad types, recyclable, non-recyclable and polluting. Recyclable waste stocks are those that can be returned to some degree as a resource, such as scrap metal and vegetable matter, but at an economic cost or by the processes of nature. Non-recyclable waste stocks are those that usually have to be buried or rendered inanimate and no longer a part of the environment, also at an economic cost. Polluting waste stocks are those that have no economic cost (subject to the laws of the day), but degrade resources or the environment in some manner. A large proportion of value lost occurs through system inefficiencies and is not included in the GDP. That which is included is counted as an output of providers whose added value *(pay and profit)* is included in the GDP, offset equally by costs deducted by users of the waste processing service.

A waste sector can be thought of primarily as a resource sector in reverse. Examples of waste processes that now occur in economic systems include water and sewage treatment, nuclear waste processing, incineration, recycling of metals, paper and other waste, land fill and sea disposal. As with resources, however, these are all technically production processes, with the objective either of abstracting some residual value, or of rendering waste safe and out of human sight. The advent of climate change has brought forward other potential processes including CO_2 sequestration. Some waste disposal systems do not provide very satisfactory solutions. Nuclear waste processing involves burying highly toxic waste. Incineration generates heat, but residual CO_2. Land fill generates methane, though this can be returned to be burnt to create heat, but additional CO_2. Most waste disposal processes in the economic domain involve a net cost. Figure 2.13 illustrates the principle of a waste stock.

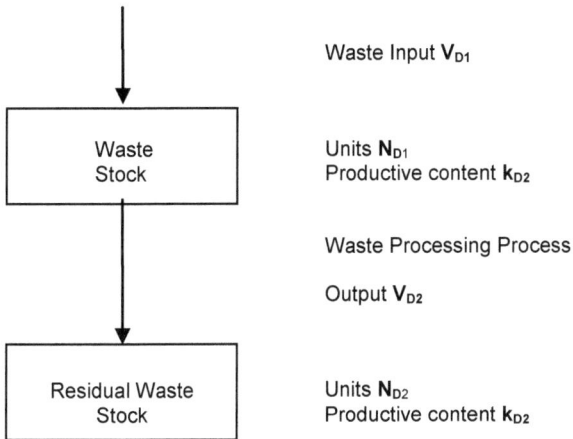

Waste Input V_{D1}

Units N_{D1}
Productive content k_{D2}

Waste Processing Process

Output V_{D2}

Units N_{D2}
Productive content k_{D2}

Figure 2.13 Waste Stock

Similar to a resource stock, waste stock throughput can be written as:

$$P_{DO}V_{DO} = P_{DO}(v_{DO}N_D) = N_D k_D (T_{DO})$$ (2.26)

Where V_{DO} is the output volume between stock 1 and stock 2 in figure 2.13, where a production process occurs.

CHAPTER 3   THERMODYNAMIC PRINCIPLES

In this chapter we set out the First and Second Laws of Thermodynamics, which are fundamental in the world of physics, and we examine the dynamics of the main processes encountered as applied to economic systems. Our initial investigations will centre on a single stock and flows immediately in and out of it, where the productive content **k** of a good remains constant, that is, it does not in any way change in shape, form, what it is made of and any other physical attributes it may have as it enters the stock, remains there or subsequently leaves it. A jelly-bean of defined size, colour and taste remains just that, though its value in economic terms can of course go up or down. Our stock can be large or small (as in 'just-in-time' processes), but the principles applying will still be the same. However, if we were to expand our analysis to include a number of stocks connected via production / consumption processes, then we would encounter both multiple values of productive content **k**, and unknown efficiency losses occurring in the production and consumption processes, which would both complicate the analysis and place restrictions on the results. The essential objective first is to set out the principles. Chapter 4 deals with production and consumption processes.

3.1   The First Law of Thermodynamics

The First Law of Thermodynamics is concerned with the principle of conservation of energy applied to systems that undergo changes of state due to transfers of heat and work across a system boundary. The First Law cannot be proved; its validity rests upon the fact that it has never been contradicted by experience. It states: *"When a closed system is taken through a cycle, the net work delivered to the surroundings is proportional to the net heat taken from the surroundings"*. For a *non-flow* gas system where equation (1.1) **PV=NkT** applies, the First Law is generally stated as:

$$Q - W = (U_2 - U_1) \tag{3.1}$$

Where **Q** is the heat passing across the boundary of the system, **W** is the work done or consumed, and **(U₂ − U₁)** is the change in internal energy arising between states 1 and 2. Imagine some gas held in a cylinder by a piston, as in figure 3.1.

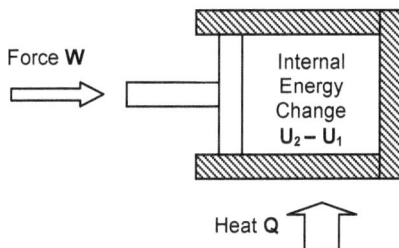

Force **W**

Internal
Energy
Change
$U_2 - U_1$

Heat **Q**

Figure 3.1 Illustration of First Law

If a force **W** is applied to the piston, the gas is compressed, reducing the volume **V** of the cylinder and raising the pressure **P**. The temperature **T** of the gas goes up, with the molecules moving around faster. The gas has therefore accumulated some internal energy ($U_2 - U_1$). Likewise, imagine the piston locked to the cylinder with a pin, holding the volume of the gas constant. Some heat **Q** from outside is applied to the cylinder. The gas gets hot, with the molecules of gas moving around faster, accumulating some internal energy ($U_2 - U_1$). The increased energy of the molecules results in an increased pressure **P** on the cylinder wall and the piston head. Equation (3.1) for the First Law applies.

Similar to equation (3.1), on a unit mass or molecule basis, as in equation (1.2) **Pv=kT**, the First Law for a *non-flow* process is generally expressed as:

$$Q - W = (u_2 - u_1) \qquad (3.2)$$

With the lower case letter **u** representing unit mass or molecule internal energy change, being generally referred to as the specific internal energy, similar to the concept of specific volume **v**.

Similar derivations for equations (3.1) and (3.2) can be set out for a thermodynamic *flow* system, though additional components of kinetic energy (the speed of flow of a mass or number of molecules along a pipe or similar), potential energy (energy by virtue of height, for example as in the height of a column of water in a dam) and an extension of internal energy to enthalpy (a wider definition of energy content) are added.

In thermodynamics it is common to consider two process types, reversible and irreversible. In a reversible process, the process is imagined to pass through a continuous series of infinitesimal equilibrium states, such that equation (3.1) can be written in a differential form:

$$dQ - dW = dU \qquad (3.3)$$

Where incremental work done **dW** is equal to pressure **P** multiplied by the incremental change in volume **dV**.

$$dW = PdV \qquad (3.4)$$

The work done for a reversible process can then be found by summing up all the increments of work. Thus:

$$W = \int_1^2 PdV \qquad (3.5)$$

In a reversible process, it is possible for the process to be gradually unwound back through the infinitesimal states to the original position, such that the pressure and volume return to their original values, and the quantities of work done are reversed. Thus for a reversible process the First Law is stated as:

$$dQ - PdV = dU \qquad (3.6)$$

Or in unit mass/molecule terms:

$$dQ - Pdv = du \qquad (3.7)$$

where **du** is the incremental change in specific internal energy.

The reversible process therefore is one that cannot be improved upon in thermodynamic terms, as it can be brought back to the starting point without loss. Figure 3.2 illustrates a reversible process.

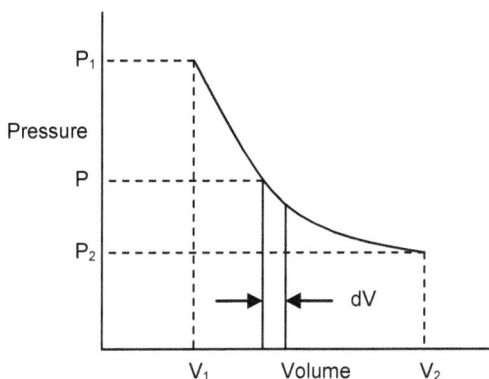

Figure 3.2 Reversible Thermodynamic Process

In an irreversible thermodynamic process, however, a complete return to the starting point would not be possible, and there would always be a difference in one of pressure or volume, if the other was returned to its original position, and there would be a net loss of potential work. In such a case the First Law of Thermodynamics is expressed as in equations (3.1) and (3.2). Figure 3.3 illustrates an irreversible thermodynamic process, in this case a loss of volume.

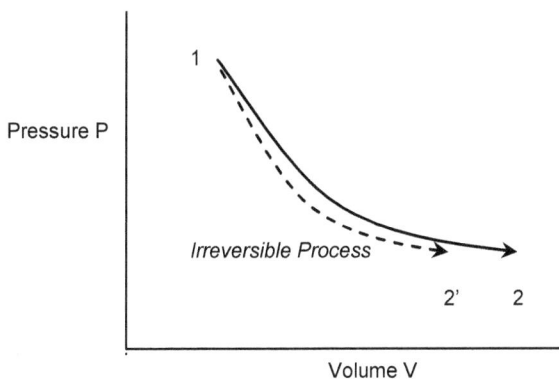

Figure 3.3 Irreversible Thermodynamic Process

For an irreversible *thermodynamic* process incremental Work Done **dW** is therefore in general less than that for a reversible process. Thus:

$$dW \leq PdV \tag{3.8}$$

Turning now to our economic system, a similar formulation to the First Law of Thermodynamics can be postulated, though, as pointed out in chapter 1, economic systems have elements of both flow and non-flow processes. We imagine a stock of a good which is fed at one end by work input value of the same good (being a function of price **P** multiplied by volume flow **V** per unit of time), with a similar work output value of the good coming out the other end of the stock, as in figure 3.4.

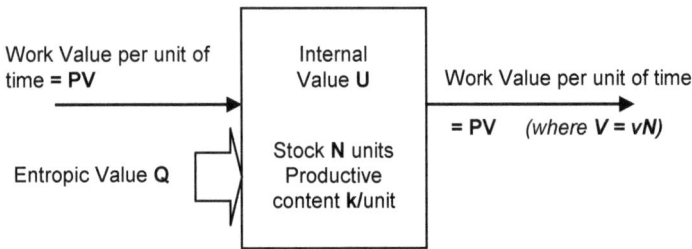

Figure 3.4 An Economic Stock

For a trader, the price of a unit leaving the stock will generally be larger than that entering it, by virtue of a profit margin, but we will not complicate matters at this point.

We imagine four events that can occur to the stock.

First, some additional or reduced *Work Done* **W** per unit of time can occur, being defined as a *change* in work value per unit of time entering and leaving the stock, by virtue of a change in volume flow of the *particular* good, *and no other good*. Thus volume flow per unit of time entering and subsequently leaving the stock could rise or fall, to meet equivalent changes in both supply and demand.

For a reversible process, incremental Work Done per unit of time would be written as $dW = PdV$, as in equation (3.4); or **Pdv** if analysing on a basis of specific volume rate **v**, as per equation (1.7). For an irreversible process, incremental Work Done **dW** would be expressed as $dW \leq PdV$, and in unit stock terms $dW \leq Pdv$, the same formats as in equation (3.8).

The concept of reversible and irreversible thermodynamic processes is not readily understood in economics. An economist might argue that by definition one cannot 'undo' a production process proceeding from a set of inputs to a set of outputs, at least not easily, and likely at a significant cost. One cannot 'undo' a loaf of bread.

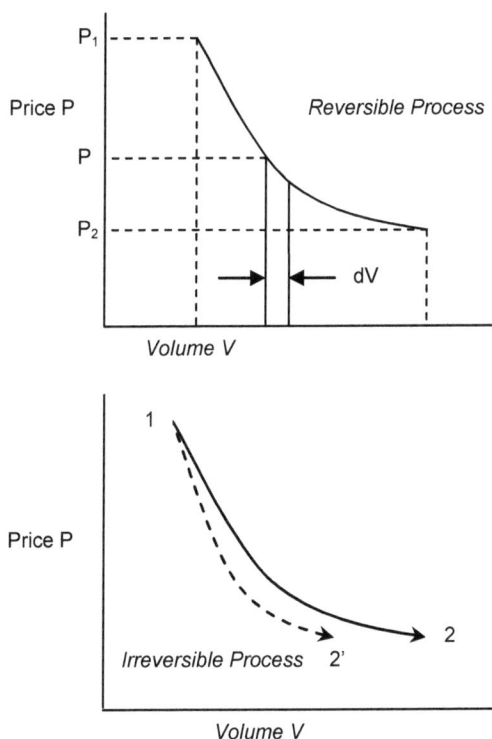

Figure 3.5 Reversible and Irreversible Economic Processes

It can be argued, however, that some degree of reversibility is achieved through the economic cycle. Thus consumption of producer capital stock in a process results in output which is circulated through the system in return for sales turnover, a part of which consists of profit which can then be used to purchase new capital stock (which has been produced elsewhere) to replace the capital stock used up. As already noted in chapter 2, this process is not a perfect one and losses will always occur. Nevertheless, the reversible process might be viewed as representing an ideal path, albeit not achieved in practice.

The second event that can occur is that value **Q** per unit of time can be put into or taken out of the system, which is *not* represented by a change in volume flow of the productive content *of the particular good*. Examples of value **Q** include a scarcity or abundance of the particular good engendered by a change in demand compared to the available supply, new money coming into the system, or the consumption of the productive content of *another different good*, some of the value of which can then be *added or transferred* to the particular good via a production process. Value **Q** therefore does *not* represent volume of throughput of the particular good gained or lost. We shall call this the *Entropic Value* added or taken out.

The third event that may occur is a change in the economic internal energy **U** of the stock, which we will call the *'Internal Value'* **U**. There is a difference between the internal value **U** and the stock productive content **Nk**. The latter is set by reference to the number of units **N** in the stock, and the *non-variable* productive content **k** of each unit. The internal value **U**, on the other hand, is a *variable* value by reference to the value of trading entering and leaving the stock per unit of time.

Imagine a trader with a stock of fashion clothes. The stock has a productive content **Nk** by virtue of the materials from which it is made, and is not a function of price. The trader is in the business of buying and selling such stock. In a good year demand is brisk and he may be able to charge higher prices for his stock. The internal value **U** of the stock is therefore perceived to go up, in tandem with the stock index of trading value **T**, even though the clothes have not changed in shape or form. If, half way through his trading year, demand suddenly collapses, the trader may be forced to sell his stock at much lower prices, and not make so many sales. The perceived internal value **U** of the stock therefore goes down.

The internal value **U** of an economic system is therefore a function of the index of trading value **T**, just as internal energy **U** in a thermodynamic system is a function of temperature **T**. The index of trading value **T**, in an economic sense, is a measure of *both* the speed at which economic stocks are being turned over *and* the relative value level (the price) being turned over.

We could further simplify the analysis by considering only a unit of stock, and write **u** = U/N, being the Specific Internal Value per unit of stock. Thence accepting that a connection exists between the specific internal value **u** and the index of trading value **T**, we could write for a single unit of the stock:

$$u = f[CT]$$ (3.9)

Where the value **C** is some function of value or productive content. Likewise, it will be recalled from equations (2.8) and (2.21) in chapter 2 that the ideal economic equation can be written as **P(vN) = NkT**, or **Pv = kT**. Thus by substitution into equation (3.9) we have:

$$C = f\left[\frac{uk}{Pv}\right]$$

Where **P** is the price of units flowing in / out of the stock and **v** is the specific volume rate. Further, it will be recalled from equation (2.2) that the specific volume rate **v** is inversely related to the lifetime ratio ξ of a unit of stock and the standard transaction time t_t ($v = 1/\xi t_t$, t_t usually a year). Thus for a standard transaction time of 1 (year) we have:

$$C = f\left[\left(\frac{u}{P}\right)\xi k\right]$$ (3.10)

which expresses the value **C** in terms of the ratio of the unit internal value to the trading price **(u/P)**, the lifetime ratio ξ and the productive content **k**. It will readily be appreciated that when the unit value **u** of a stock is imputed as being equal to its unit trading price **P**, then the value **C** would appear to be reduced to some function of its productive content **k** and its relative lifetime ξ in the stock. For the moment, therefore, we could express the value **C** as being equal to:

$$C = f[\xi k]$$ (3.11)

We will return to consideration of the value **C** in more depth after a discussion of the Second Law of Thermodynamics. Suffice to say at this point in time that the value **C** in thermodynamic terms is usually called the Specific Heat. In economic terms we will call this the *Specific Value*.

The fourth event that can occur is that the number of stock units **N** may not remain constant. For instance, industrial stock quantities tend to increase in size as production flow increases. In such cases, there is a difference between the flow of inputs and outputs to the stock, and it might be preferable therefore to work in terms of the shortened ideal economic equation (1.5) **Pv=kT,** where **v=V/N,** or other method, to accommodate the variance.

Thus in economic terms, combining all of the above, according to the First Law of Thermodynamics, a change in the internal value $(U_2 - U_1)$ will be equal to the addition of work done/consumed **W** and entropic value **Q** entering or leaving the system, depending upon the directions of flow.

The concept of change in internal value is one that can incorporate changes in both volume flow of productive content and entropic value. The following examples are illustrative of the process.

First, we suppose that a seller has a stock of **N** units with productive content **k** each. The seller makes sales of **V** units per year at price **P**. The unit stock turnover is given by the equation **PV=NkT,** as per equation (1.3) chapter 1, where **T** is the index of trading value of the stock. We now suppose that the buyer is prepared to pay an additional amount **ΔQ** per annum for the same unit output, so as to assure his supply against other buyers. The seller is therefore richer by this additional amount and can for example raise his price **P** to **(P+ΔP)** to match the additional monies. Thus for the same input, the internal value of the seller's stock has risen by **ΔU=ΔQ**, and his index of trading value has risen to **(T+ΔT),** in similar proportion to the rise in price. Thus **ΔT/T = ΔP/P.** It will be noted in this example that no change in volume throughput has occurred.

As a second example, suppose that instead of the buyer presenting the seller with a present of **ΔQ** per annum, a donor (or a bank) lends the seller an amount **ΔQ** to do with as he pleases, so long as he keeps the money in the business. He can choose to increase his volume flow rate of stock purchases to a rate of **(V + ΔV)**, and hope to sell it onwards at the same price **P**, thus increasing his work output by **ΔW** (equals **PΔV**). The internal value of the stock rises by **ΔU**, with a rise in the index of trading value of **ΔT/T = ΔV/V.** It will be noted in this example that no change in output price has occurred.

As a third example, a manufacturer may have facilities to consume additional amounts of *another input (labour, resources etc)* which can be converted via the production process to additional throughput per unit of time of the particular good. There is thus a conversion first of productive content of inputs to value **Q** (net of efficiency losses), and from value **Q** to Work Done **W** on the particular good, accompanied by a possible change in internal value.

The essential points arising from the above development are first, that entropic value **Q,** entering a stock system of a particular good, does not represent a volume flow of productive content of the particular good; it is essentially money, the promise of money, a difference in asset values, a change in preferences, or a value arising from the consumption of *other* goods. Second, the internal value **U** of a stock of a particular good is a measure of the stock value that can be turned around at a certain rate or can be added to, and a rise in internal value can therefore occasion a rise in price, a rise in volume flow or both. Third, there is good reason to call the property **T** an index of trading value, and not just a velocity of circulation, since changes in the index can represent a change in volume flow, a change in price, or both.

It should be strongly emphasised yet again that by stating this we are *not* implying that an absolute scale of monetary 'productive content' can be constructed for a currency, in the manner of weight or energy content. Economics is very much a comparative discipline, and values of currency can and do change. Chapter 5 deals with the application of thermodynamic principles to a monetary stock.

3.2   The Second Law of Thermodynamics

There is nothing implicit in the First Law of Thermodynamics to say that some proportion of heat supplied to an engine must be rejected, and therefore that the cycle efficiency cannot be unity. All that the First Law states is that net work cannot be produced during a cycle without *some* supply of heat, i.e. that a perpetual motion machine of the first kind is impossible.

Likewise in our economic system net work output cannot be achieved without a supply of value arising from input of resources and consumption of some human effort and/or capital stock.

The Second Law of Thermodynamics however is an expression of the fact that some heat must *always* be rejected during a cycle. The law can be stated as: *"It is impossible to construct a system which will operate in a cycle, extract heat from a reservoir, and do an equivalent amount of work on the surroundings."* There is always some heat left over that cannot be converted into work output. This fact was discussed at length in chapter 2.

In economic terms the law could be stated as: *"It is impossible to construct an economic system which will operate in a cycle, extract productive content from a reservoir and do an equivalent amount of work, in terms of productive content, on the surroundings."* There is always a bit of productive content left over that cannot be incorporated into product output. Figure 2.6 in chapter 2 illustrates this point. Input value Q_1 from resources is consumed to produce output W, with some waste Q_2 left over. Thus in the cycle, the first law is stated as:

$$W = Q_1 - Q_2 \qquad (3.12)$$

And the cycle efficiency is stated as:

$$\eta = \frac{Work\ Done}{Value\ Supplied} = \frac{W}{Q_1} = \frac{Q_1 - Q_2}{Q_1}$$

$$\eta = 1 - \frac{Q_2}{Q_1} \qquad (3.13)$$

The work of Ayres and Warr, referred to at chapter 2, showed that the efficiency of exergy processes relating to energy resources was of the order of 15%, with an overall loss of useful work of 85%.

However, the problem with the structure of an economic system is that it appears to defy the Second Law of Thermodynamics. This apparent non-sequitur arises because no subtraction is made for efficiency losses; everything is calculated on the basis of a valuation being placed only on the final output of each product and service. Nevertheless, despite the apparent difference between the productive content of resources and the economic value attached to them by humankind, it is reasonable to assume that the economic value added attached at each stage to final output, is distributed in proportion to the final productive content of the goods included in the economic system.

From the specific stock process set out in this section, it will be noted that although this does not involve a change in productive content k, the example of the fashion trader showed that changes in the three factors of work done W, internal value U and entropic value Q do have an impact on each other. It is also well-known in economic analysis that the inter-relationship of price to volume in a process is dependent on the characteristics of supply and demand. It is therefore necessary to pursue thermodynamic analysis further in order to set this in context.

Now in a closed reversible thermodynamic system there exists a property **S**, such that a change in its value between two states is equal to:

$$S_2 - S_1 = \int_1^2 \left(\frac{dQ}{T} \right)_{rev} \tag{3.14}$$

Or in differential form:

$$dS = \left(\frac{dQ}{T} \right)_{rev} \quad \text{or} \quad dQ = TdS_{rev} \tag{3.15}$$

For the unit stock format **Pv=kT**, this would be written as **Tds**, using lower case.

The property **S** is called the *Entropy* of the system, and the value **dS** is the incremental change in entropy. The suffix 'rev' is added as a reminder that the relation holds only for a reversible process. The reader will readily note from the above that entropy change is a function of the entropic value change **ΔQ** added or taken away, not represented by a change in volume flow of productive content, which we discussed earlier

In thermodynamics, entropy is a property that measures the amount of energy in a physical system that cannot be used to do work. In statistical mechanics it is defined as a measure of the probability that a system would be in such a state, which is usually referred to as the "disorder" or "randomness" present in a system. Given that systems are not in general reversible then, following whatever means are applied to return a system to its starting point, the net change in cycle entropy is commonly stated as:

$$\oint \frac{dQ}{T} \geq 0 \tag{3.16}$$

Thus to repeat the Second Law; it is impossible to construct a system which will operate in a cycle, extract heat from a reservoir and do an equivalent amount of work on the surroundings. Entropy tends to rise. It is a measure of dispersed value.

Now, by combining equation (3.6) for the First law and (3.15) for the Second law and inserting the term for the incremental work done $\mathbf{dW} = \mathbf{P}dV$, we could also construct an entropy function for an economic system:

$$TdS = dU + PdV \qquad (3.17)$$

And in unit stock terms $(\mathbf{N=1})$ we can write:

$$Tds = du + Pdv \qquad (3.18)$$

Equations (3.17) and (3.18) set out the general relations between the properties and, when integrated, give the change in entropy occurring between two equilibrium states for a reversible process. It should be noted that entropy change in economic terms is associated with changes in *flow* of economic value.

A series of economic processes is now examined to develop what the concepts mean in economic terms. The key relationships between price and volume flow are illustrated at figure 3.6.

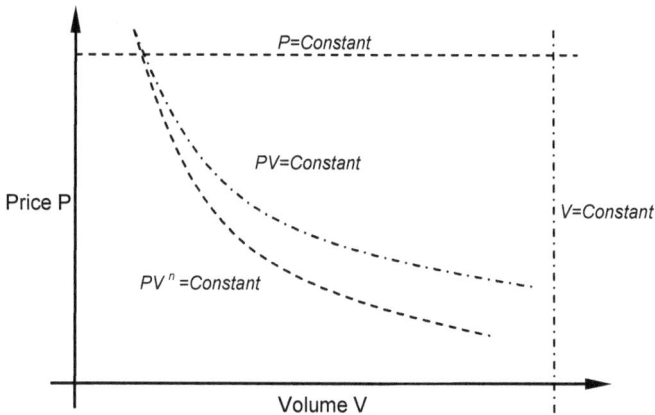

Figure 3.6 Price – volume relationships

3.3   Constant Volume Process

By definition, a constant volume process is one involving no change in volume flow **V** per unit of time, with volume flow in an out of a stock being constant. The work done **PdV** to increase or decrease volume flow through the economic system is therefore zero. From equation (3.17) for a reversible process we have:

$$TdS = dU + PdV$$

Thence, as **PdV** is zero:

$$TdS = dU \tag{3.19}$$

Thus a change in the entropy **dS** to the stock system is reflected only as a change in internal value **dU**. Now by differentiating the ideal economic equation **PV = NkT** we can also write:

$$PdV + VdP = NkdT \tag{3.20}$$

And remembering in this case that volume change is zero **dV=0**, then:

$$VdP = NkdT$$

Hence by substituting in **PV =NkT** again we have:

$$\frac{dP}{P} = \frac{dT}{T} \tag{3.21}$$

And:

$$\frac{P_2}{P_1} = \frac{T_2}{T_1} \tag{3.22}$$

Thus the percent change in the price of output flow in the process (the volume flow rate not having changed) changes exactly in proportion to the percent change in the index of trading value **T**, arising from the input or output of entropic value **dQ = TdS** to and from the system. Nothing has been done to the items in the system, no work has been done; they are just perceived by the players in the system as having more or less value, by virtue of the entropic value **dQ** introduced or taken away. Economists might indicate that a change in price/value of this kind could arise from changes in scarcity, abundance or utility values, or the difference between supply and demand.

Now in order to compute the change in entropy associated with this process, we have first to set out a relationship between the change in the internal value **dU** and the change in the index of trading value **dT**. Similar in structure to equation (3.9), we could write for a single stock unit and for a multiple unit stock:

$$du = C_v dT \quad \text{and} \quad dU = NC_v dT \quad (3.23)$$

where C_v is a constant (for an 'ideal' economic system), which we shall call the *Specific Value at constant volume,* being analogous to the specific value **C** developed in equation (3.11).

The thermodynamic analogy here is the specific heat at constant volume, being the heat required to raise the temperature of a unit of a gas system by one degree of a scale of temperature. The specific heat of a gas is commonly computed in thermodynamics terms by reference to either unit mass or quantity. The usual measure of the latter is per mole. For a monatomic ideal gas the specific heat at constant volume $C_v = (3/2) N_A k$, where N_A is Avogadro's Number. Thus specific heat is measured as heat value relating to a multiple of numbers of molecules.

In this book the *Specific Value* of a good at constant volume C_v in an economic system is defined as the amount of value **du** required to be introduced to the internal value of a unit of stock to change the index of trading value by **dT**, but without any net change in volume flow in or out of the system. It is a measure of ability to store the entropic value that is introduced by **dQ**. In economic terms utility has risen or declined, but nothing of substance has been added or taken away.

It might be supposed that the value C_v for an economic stock good would be a constant. However, in gas systems, according to the kinetic theory of gases, the specific heat at constant volume is actually dependent upon the complexity of the gas molecules. A simple molecule requires less energy to increase its momentum and raise its temperature, than does a complex one, according to the number of *'degrees of freedom'* – dimensional, rotational and vibrational energies (quantum mechanics introduces yet further degrees of freedom, those of electronic and nuclear). And in reverse, a complex molecule releases more energy for a given drop in temperature than does a simple molecule. The question therefore arises therefore as to whether such a variation is possible in an economic system.

The answer proposed in this book is that variation in C_v is likely to depend on the 'complexity' of a good in terns of its attributes. In economic terms *'degree of value'* might be a better description than *'degree of freedom'*. To illustrate the concept, money in the form of cash clearly has a nominal value as a means of exchange, with a relatively short lifetime. It might be deemed to have a low *degree of value* C_v. By contrast, both goods that are produced and income generating securities are more complex, containing productive content that can only be released over time.

The two obvious connections therefore, as hinted at equation (3.11), are that the specific value C_v is a function of both the productive content k of the particular good, and the effective lifetime of the good, notated as ξ at equations (2.2) and (3.11). However, some goods, such as gold and diamonds have aesthetic value (to humans), adding complexity to the valuation. To allow for additional complexity, we choose to define the specific value at constant volume C_v to be proportional to the embodied value/productive content k, and another factor ω, which will encompass both the lifetime and other aspects. Hence in our constant volume economic system we could write:

$$C_v = \omega k \tag{3.24}$$

Where ω might be called the *Value Capacity Coefficient*.

In thermodynamics, the specific heat of a gas is determined by means of highly controlled experiments, to ensure that no heat losses and other factors occur to nullify the results. In economics of course, all values being relative, designing an experiment is likely to be that much more difficult to do. Nevertheless, stating the specific value at constant volume C_v in this way enables us to continue with the analysis, even if a value cannot immediately be attached – though a starting point might be the effective lifetime or turn-round time of a stock.

Now by combining equations (3.15), (3.19), (3.23) and (3.24), the entropy change in the economic process is then written as:

$$dS = \left(\frac{dQ}{T}\right)_{rev} = \left(\frac{dU}{T}\right)_{rev} = NC_v\left(\frac{dT}{T}\right)_{rev} = N\omega k\left(\frac{dT}{T}\right)_{rev} \tag{3.25}$$

Thence by integrating we have, for a reversible process:

$$S_2 - S_1 = N\omega k \ln\left(\frac{T_2}{T_1}\right) \qquad (3.26)$$

And by substituting in equation (3.22) we have:

$$S_2 - S_1 = N\omega k \ln\left(\frac{P_2}{P_1}\right) \qquad (3.27)$$

which equates the change in entropy in terms of the value capacity coefficient ω, and the changes in the index of trading value **T** and price **P**.

In differential form equations (3.26) and (3.27) can be written as:

$$dS = N\omega k\left(\frac{dP}{P}\right)_{rev} = N\omega k\left(\frac{dT}{T}\right)_{rev} \qquad (3.28)$$

Thus entropy change in the constant volume process is proportional to the percent change in price or index of trading value. Figure 3.7 illustrates the Constant Volume process, in terms of a **P-V** diagram and an **S-T** diagram.

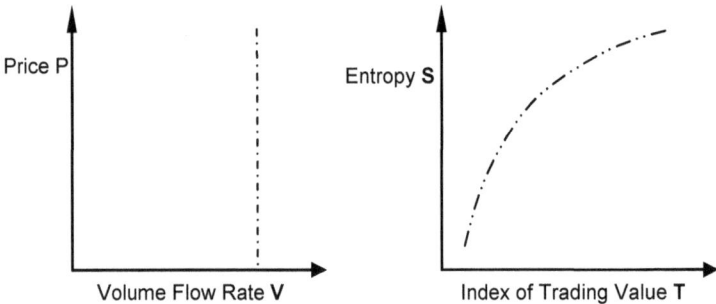

Figure 3.7 Constant Volume Process

By definition, a process that is close to a constant volume process is highly *inelastic*, as a price change can only result in a small change in volume.

3.4   Constant Price Process

In contrast to the constant volume process, a constant price process is one than involves a change in volume flow per unit of time, but no change in price. Work done is *not* therefore zero, and any entropic value **dQ** entering or leaving the system must equate to the work done **dW** in changing the volume flow, plus the change in the internal value **dU** of the system. Hence equation (3.17) for a reversible process is stated as:

$$TdS = dU + PdV \tag{3.29}$$

As with the constant volume process, by differentiating the ideal economic equation **PV=NkT** we have:

$$PdV + VdP = NkdT$$

But, since in this process price remains constant, **VdP** is therefore zero and we can then write:

$$PdV = NkdT \tag{3.30}$$

Hence by combining equations (3.23), (3.29) and (3.30) we have:

$$
\begin{aligned}
TdS &= NC_v dT + PdV \\
&= NC_v dT + NkdT \\
&= NC_p dT
\end{aligned}
\tag{3.31}
$$

Where $C_p = (C_v + k)$ is a constant, which we shall call the *Specific Value at constant price*, being analogous to the specific heat at constant pressure in a thermodynamic system, in a similar manner to the constant volume process discussed earlier.

It will be recalled also from equation (3.24) that the specific value at constant volume using our value capacity coefficient was $C_v = \omega k$; thence we could write for a constant price process:

$$
\begin{aligned}
C_p &= \omega k + k \\
&= (\omega + 1)k
\end{aligned}
\tag{3.32}
$$

The higher value of the specific value C_p at constant price, compared to that of the specific value C_v at constant volume, recognises that in adding value to the internal value U, volume movement of units takes place. Additional value is flowing through, i.e. not only the entropic value ωk (equation 3.28), but also volume of real productive content k into and out of the stock.

Now by substituting the ideal equation $PV = NkT$ back into equation (3.30) and remembering that price is constant we have:

$$\frac{dV}{V} = \frac{dT}{T} \tag{3.33}$$

And:

$$\frac{V_2}{V_1} = \frac{T_2}{T_1} \tag{3.34}$$

Thus the percent change in the volume flow rate V matches the percent change in the index of trading value T; which is what one might expect for a constant price process. A change in the index of trading value finds its way wholly into a change in volume flow, and not price.

Similarly by combining equations (3.15), (3.31) and (3.32), the entropy gain for the process is written as:

$$dS = \left(\frac{dQ}{T}\right)_{rev} = NC_p\left(\frac{dT}{T}\right)_{rev} = Nk(\omega+1)\left(\frac{dT}{T}\right)_{rev} \tag{3.35}$$

Thence by integrating we have for a constant price reversible process:

$$S_2 - S_1 = Nk(\omega+1)\ln\left(\frac{T_2}{T_1}\right) \tag{3.36}$$

And by substituting in equation (3.34) we have:

$$S_2 - S_1 = Nk(\omega+1)\ln\left(\frac{V_2}{V_1}\right) \tag{3.37}$$

Thus stating the change in entropy for the constant price process in terms of the changes in the index of trading value and the associated volume flow rate. Figure 3.8 illustrates the Constant Price process, in terms of a **P-V** diagram and an **S-T** diagram.

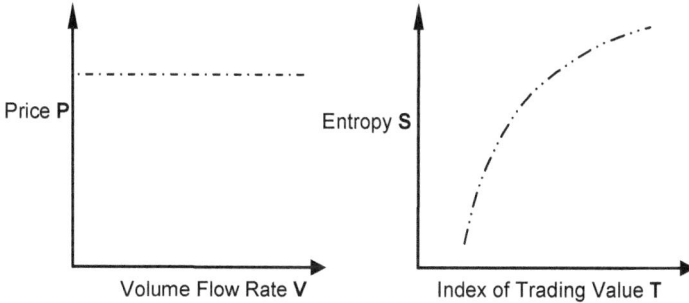

Figure 3.8 Constant Price Process

By definition, a process that is close to a constant price process is highly *elastic*, as a small change in price can result in a large change in volume flow.

3.5 Iso-trading Process

As its name suggests, the iso-trading process is one where no change in the index of trading value occurs, that is **dT = 0**. The equivalent thermodynamic process is the isothermal case where temperature change is zero. In mathematical terms we can write:

$$PV = Z \qquad (3.38)$$

where **Z** is a constant, and price varies *inversely* with volume flow. The greater the flow, the lower the price has to be in order for no change in the index of trading value to take place. This formula has common usage in standard economics textbooks on demand curves, and the shape of the curve is depicted at figure (3.7).

In our model, since the index of trading value **T** is constant, there is no change in incremental internal value **dU** in the stock, and therefore any change in incremental work done **dW** is reflected as a change in entropic value **dQ**. Thus for a reversible process we have:

$$dQ = dW = PdV$$

$$dQ = PdV \qquad (3.39)$$

And differentiating the ideal economic equation **PV=NkT**, we have:

$$PdV + VdP = NkdT$$

And since **dT=0**, we have:

$$PdV + VdP = 0$$

Hence:

$$\frac{dP}{P} = -\frac{dV}{V} \qquad (3.40)$$

Indicating, as would be expected, that a percent change in price is balanced by an opposite percent change in the volume flow rate.

By substituting equation (3.39) into equation (3.15) for the entropy change we have:

$$dS = \left(\frac{dQ}{T}\right)_{rev} = \frac{1}{T}\left(PdV\right)_{rev} \qquad (3.41)$$

And by further substituting in **PV=NkT**, we have:

$$dS = Nk\left(\frac{dV}{V}\right)_{rev} \qquad (3.42)$$

And

$$dS = -Nk\left(\frac{dP}{P}\right)_{rev} \qquad (3.43)$$

Hence by integrating we have for a reversible process:

$$S_2 - S_1 = Nk \ln\left(\frac{V_2}{V_1}\right) \qquad (3.44)$$

And:

$$S_2 - S_1 = Nk \ln\left(\frac{P_1}{P_2}\right) \tag{3.45}$$

Thus we have a logarithmic relationship of entropy change ΔS with change in volume flow rate V, and an equal and opposite logarithmic relation with change in price P. Figure 3.9 illustrates the Iso-trading process, in terms of a P-V diagram and an S-T diagram.

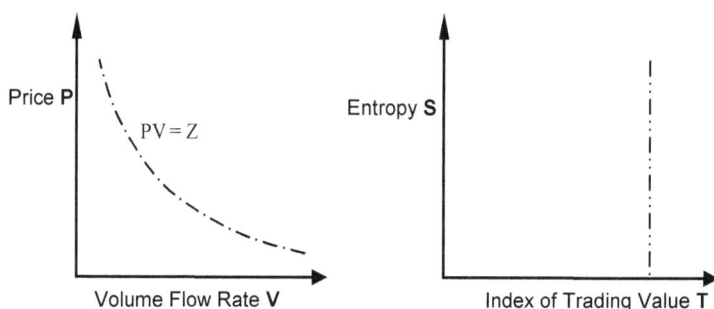

Figure 3.9 Iso-trading Process

By definition, an iso-trading process is one that is neither highly *elastic* nor highly *inelastic*, as a percent change in price engenders a similar percent change in volume demand.

3.6   Polytropic Process

A more general type of relationship of price against volume found in economic processes is of the form:

$$PV^n = Z \tag{3.46}$$

Where Z is a constant and n is a factor known as the *elastic* index. This is easily confirmed by differentiating the equation to give:

$$\frac{dP}{P} = -n\left(\frac{dV}{V}\right) \tag{3.47}$$

Supply and demand curves are often drawn to this formula with demand curves having a positive value of **n** and supply curves a negative value. In thermodynamics such processes are called *Polytropic* processes and we shall use the same term here. It will noted that when **n = 0** the relationship reduces to a constant price process, and when **n = ∞** it reduces to a constant volume one.

Now, referring back to our formula for the work done **W**, substituting in **PV^n=Z**, we have for a reversible process:

$$W = \int_1^2 PdV = \int_1^2 \frac{Z}{V^n}dV$$

Thence by integration and substitution we obtain:

$$W = \left(\frac{P_2V_2 - P_1V_1}{1-n}\right) \tag{3.48}$$

And by further substitution of the ideal economic equation **PV=NkT**:

$$W = N\left(\frac{k}{1-n}\right)(T_2 - T_1) \tag{3.49}$$

Substituting the above back into equation (3.1) for the First Law relating entropic value to work done and the change in internal value we obtain:

$$Q - N\left(\frac{k}{1-n}\right)(T_2 - T_1) = U_2 - U_1 \tag{3.50}$$

And re-arranging and substituting in the integrated form of equation (3.23) for the internal value, and equation (3.24) for the specific value at constant volume:

$$Q = NC_v\left(T_2 - T_1\right) + N\left(\frac{k}{1-n}\right)\left(T_2 - T_1\right)$$

$$= Nk\left(\omega + \frac{1}{1-n}\right)\left(T_2 - T_1\right) \tag{3.51}$$

Finally we have an expression for the change in entropy in terms of the index of trading value T:

$$S_2 - S_1 = Nk\left(\omega + \frac{1}{1-n}\right)\ln\left(\frac{T_2}{T_1}\right) \tag{3.52}$$

This equation can also be re-stated in terms of changes in price and changes in volume flow, by substituting in $PV^n = Z$, although we will not clutter up the picture here. There are nevertheless three expressions setting out the relationships between volume flow, price and the index of trading value:

$$\frac{P_2}{P_1} = \left(\frac{V_1}{V_2}\right)^n \tag{3.53}$$

$$\frac{T_2}{T_1} = \left(\frac{P_2}{P_1}\right)^{\frac{n-1}{n}} \tag{3.54}$$

$$\frac{T_2}{T_1} = \left(\frac{V_2}{V_1}\right)^{1-n} \tag{3.55}$$

Figure 3.10 illustrates the polytropic process, in terms of a P-V diagram and an S-T diagram:

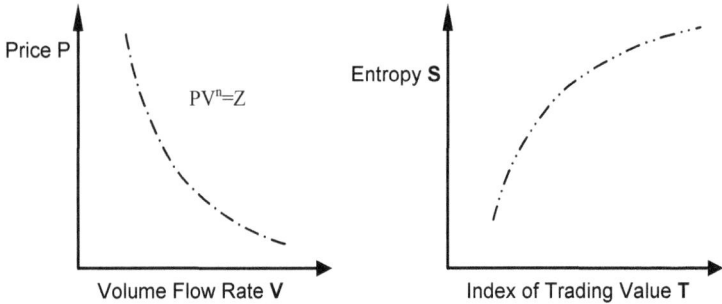

Figure 3.10 Polytropic Process

3.7 Isentropic Process

An important special case of the Polytropic process is the *Isentropic* case, where incremental entropy change **dS** is zero, with no entropic value **Q** entering or leaving the system. Thus in differential form for a reversible process, the incremental work done **dW** is equal and opposite to the change in internal value **dU**:

$$dW = -dU = -NC_v dT \tag{3.56}$$

Substituting in **PdV** for **dW** and setting alongside the ideal economic equation **PV=NkT** we have:

$$PdV = -NC_v dT \qquad \text{(First Law)}$$

$$NkdT = PdV + VdP \qquad \text{(Ideal Economic Equation)}$$

Eliminating **dT** from these equations and re-arranging we obtain:

$$0 = \left(1 + \frac{C_v}{k}\right)PdV + \left(\frac{C_v}{k}\right)VdP$$

And since $C_p = (C_v + k)$, this reduces to:

$$0 = C_p PdV + C_v VdP$$

By writing $C_p/C_v = \gamma = (\omega+1)/\omega$ this then becomes:

$$\gamma \frac{dV}{V} + \frac{dP}{P} = 0$$

And finally by integrating we get:

$$PV^\gamma = Z \qquad (3.57)$$

Which is another form of equation (3.46) for the Polytropic process, with the elastic index $\gamma = (\omega+1)/\omega$ being a function of the value capacity coefficient ω.

We can therefore substitute in γ for n to arrive at the isentropic relationships between volume flow, price and the index of trading value:

$$\frac{P_2}{P_1} = \left(\frac{V_1}{V_2}\right)^\gamma \qquad (3.58)$$

$$\frac{T_2}{T_1} = \left(\frac{P_2}{P_1}\right)^{\frac{\gamma-1}{\gamma}} \qquad (3.59)$$

$$\frac{T_2}{T_1} = \left(\frac{V_2}{V_1}\right)^{1-\gamma} \qquad (3.60)$$

Since by definition there is no change in entropy in this process, all value changes to the internal value of the system involve only changes in real volume flow, with no change in entropic value. Figure 3.11 illustrates the Isentropic process, in terms of a **P-V** diagram and an **S-T** diagram.

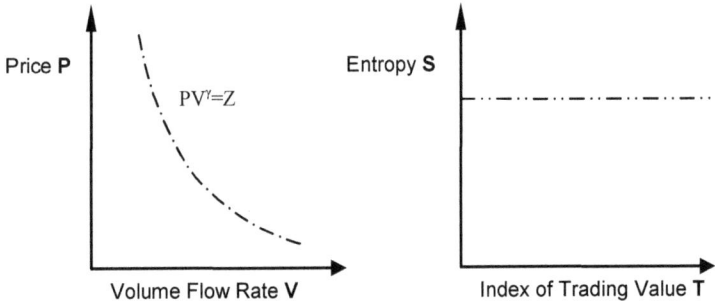

Figure 3.11 Isentropic Process

3.8 Process Entropy

The equations for entropic gain in the preceding reversible processes examined all have a common form, and can all be derived from the entropic gain for the Polytropic process. It will be recalled from equation (3.52) that the expression for the change in entropy was derived as:

$$S_2 - S_1 = Nk\left(\omega + \frac{1}{1-n}\right)\ln\left(\frac{T_2}{T_1}\right) \qquad (3.61)$$

This can be re-stated as:

$$S_2 - S_1 = Nk\lambda \ln\left(\frac{T_2}{T_1}\right) \qquad (3.62)$$

Where

$$\lambda = \left(\omega + \frac{1}{1-n}\right) \qquad (3.63)$$

may be called the *Entropic Index*.

Hence the entropy change for a given process is related to both the change in the index of trading value, and the value of the entropic index λ. The latter is a function only of the value capacity coefficient ω and of the elastic index **n** of the process. Figure 3.12 shows how the entropic index varies with changes in the elastic index.

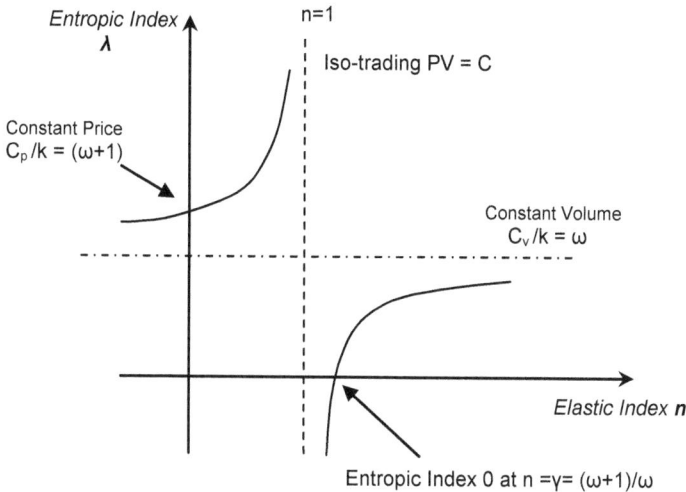

Figure 3.12 Relationship between Entropic Index and Elastic Index

At the point where the elastic index **n** is equal to $(\omega+1)/\omega)$, the entropic index becomes zero with no change in entropy occurring. This equates to the isentropic case (section 3.7). At the vertical line where the elastic index **n** is equal to 1, we have the iso-trading process (section 3.5) where the entropy change is not related to change in the index of trading value, but only to changes in price and volume flow. At the horizontal line where the entropic index is equal to ω we have the constant volume process (section 3.3). Last, we have that the entropic index λ for a constant price process is equal to $(\omega+1)$ (section 3.4), which implies an elastic index **n** equal to zero, with price not a function of volume flow.

It can be seen that as the elastic index approaches unity, the curves swing wildly up and down, entailing high positive and negative values of entropic index. Where the line crosses the x-axis the entropic index λ becomes zero. This occurs when the elastic index attains the particular value **n=γ**, and is a function of the value capacity coefficient ω of the particular good. Figure 3.13 illustrates some values of the entropic index λ set against both the elastic index **n** and the value capacity coefficient ω.

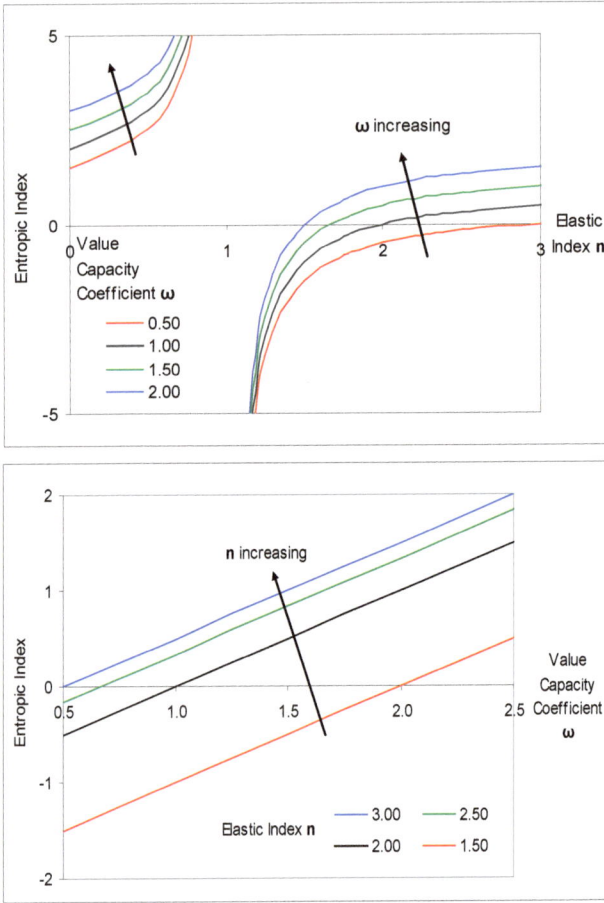

Figure 3.13 Entropic Index, Elastic Index and Value Capacity Coefficient

A further way of looking at the process is to consider deviations of the entropic index λ from the isentropic position. It will be recalled from equation (3.63) and figure 3.12 that the condition of nil entropy gain is satisfied when the elastic index n is equal to $\gamma = (\omega+1)/\omega$. Figure 3.14 sets out a graph of elastic index versus the value capacity coefficient ω. The line on the graph represents a locus of points of nil entropy gain where $n = \gamma = (\omega+1)/\omega$.

It can be seen that the isentropic elastic index γ is low for high values of value capacity coefficient ω, and is high for low values of value capacity coefficient ω. High values of elastic index are associated with an inelastic position, whereby output volume is not effected much by changes in price; and vice versa.

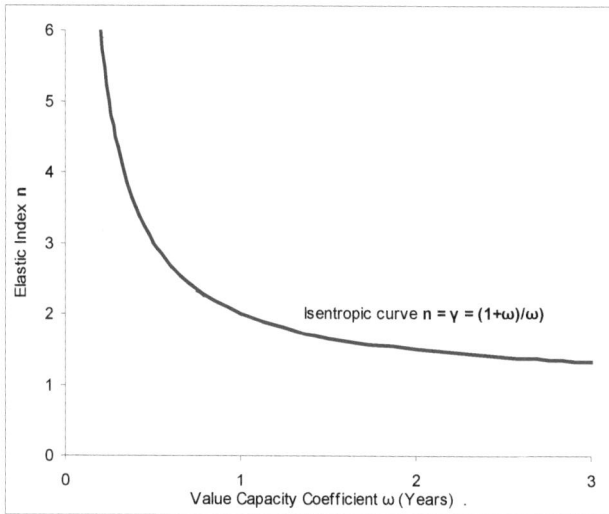

Figure 3.14 Elastic Index as a function of Value Capacity Coefficient

The position for assets and financial instruments such as bonds, with long lifetimes and maturity dates further to the right along the X-axis is therefore different from that for short-term economic assets and instruments such as cash.

A process where the curve shifts either side of the isentropic position, with elastic index **n** greater than or less than the isentropic elastic index γ, may change the entropic index λ and thence reduce or amplify changes in the index of trading value **dT/T**. This point is important when we come to consider a monetary model at chapter 5.

From all of the above, it can be seen that the entropy gain for the general polytropic process is of a logarithmic form, as per the general equation (3.61). This is augmented both by the value capacity coefficient ω and the elastic index **n**, as shown in figure 3.15.

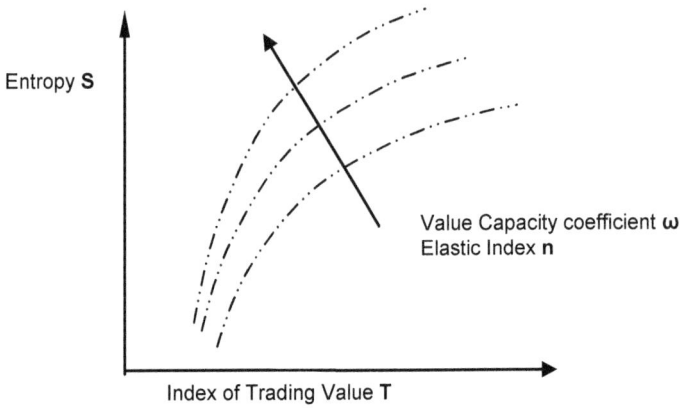

Figure 3.15Entropy for a Polytropic Process

Finally, it is of interest to note that by algebraic manipulation of equations (3.54), (3.55) and (3.62) the incremental entropy change for the polytropic process can be expressed in three forms:

$$ds = k\left(\omega + \frac{1}{1-n}\right)\frac{dT}{T} \qquad (3.64)$$

$$ds = k(\omega - \omega n + 1)\frac{dV}{V} \qquad (3.65)$$

$$ds = k\left(\frac{-1}{n}\right)(\omega - \omega n + 1)\frac{dP}{P} \qquad (3.66)$$

Thus incremental entropy change **ds** is equated to factors multiplied by percent changes in the index of trading value, the volume flow rate, or the price. Thus, in terms of economic entities, incremental entropy change **ds** is effectively similar in concept to percent rates for growth, inflation and interest. This property should be borne in mind when reading later chapters of this book.

3.9 Thermodynamics and Utility

A number of researchers have highlighted similarities between economic utility theory and thermodynamic concepts, in particular entropy. Candeal, Miguel et al (2001) describe a similarity between the utility representation problem in utility theory and the entropy representation problem related to the second Law of Thermodynamics. Sousa and Domingos (2005, 2006) describe a number of aspects of both utility theory and thermodynamics. Smith & Foley (2002, 2004) highlight similarities between utility and entropy. From the theory developed in this chapter some similarities between utility and entropy also occur, perhaps being even more strikingly apparent, first met with the Constant Volume process.

Before proceeding further, it should be noted that in economics it is common to use the character **U** to signify utility. However, since the same character is used in thermodynamics to signify the internal energy of a gas, which we have used also here in connection with the internal value held in a stock by virtue of its throughput; to avoid confusion we will change the economic utility character to **Y**.

The economic concept of utility **Y** is one that is not readily understood and appreciated by scientists, and some basic explanation is necessary to provide a bridge between the two disciplines. In simple terms, utility theory posits that consumers can choose between consuming more or less of a variety of goods, according to their particular circumstances and preferences. Consumers are said subconsciously and individually to attach particular utility values to each good, and to endeavour to maximise their total utility within an overall income/budget constraint. Economists use such constructs as indifference curve maps and Edgeworth boxes to show how one consumer's preferences are reconciled to another's. The total utility **Y** for a range of goods **(a....n)** for a particular consumer can be expressed as:

$$Y = f\left(V_a, V_b....V_n\right)$$

subject to $\left(P_a V_a + P_b V_b + ...P_n V_n\right) \leq m$ (3.67)

Where **V** is volume, **P** is price and **m** is a budget value constraint. Utility, in economic parlance, is therefore couched in terms of a consumption set of volumes of different goods, within an overall constraint of $\sum \mathbf{PV} \leq \mathbf{m}$.

A basic tenet of utility theory is that as consumers consume more of a particular good, their utility with respect to that good rises, but grows slower and slower as their level of satisfaction causes them to turn to other possibilities on which to spend their income. Thus their marginal utility with respect to the particular good falls with increasing volume.

Proceeding further, the Law of Diminishing Marginal Utility states that at consumer equilibrium the marginal utility of one good with respect to volume $\partial Y/\partial V$ divided by its price P is equal to the marginal utility of another good with respect to volume change divided by its price:

$$\frac{\left(\partial Y/\partial V\right)_a}{P_a} = \frac{\left(\partial Y/\partial V\right)_b}{P_b} = ... = \frac{\left(\partial Y/\partial V\right)_n}{P_n} \tag{3.68}$$

The equation is set out in terms of partial utility derivatives for each good, within the total utility and the budget constraint at equation (3.64). The law constitutes a basic input to the law of downward sloping demand curves – price being inversely a function of volume, as per most of the curves in figure 3.7 and subsequently developed in this chapter.

A number of utility functions are in common use in economics to explain particular preferences, including constant elasticity of substitution, isoelastic, Cobb-Douglas, exponential and linear. The isoelastic case for instance is of the form $Y=f(X)=(X)^{1-a}/1-a$, which, for a value of $a=1$, in the limit reduces to $Y=\ln(X)$, the familiar log curve often used in illustrations of utility.

A scientist will readily appreciate that in equation (3.68) no direct reference is made to the underlying productive content/embodied value, energy or other scientific measure for each good, only the price P and the marginal utility values with respect to volume $\partial Y/\partial V$ attached to them by a particular consumer. These are both variables, and their value, according to economic theory, depends upon the view of each consumer and supply and demand only. Two points to note here therefore are first, that the utility and price attached to any one good can vary compared to the productive content of the good, which does not vary, and second, consumers view the productive content as being *net* of any exergy and efficiency losses incurred in its production which do not appear in the economic balance sheet.

We have therefore to investigate whether a link between utility theory and thermodynamics can be established, using the theoretic structure set out so far in this chapter, and what differences may exist between the two disciplines.

First, it should be noted that immediately before a choice decision, our consumer has not fulfilled his utility; and this remains unsatisfied, free and positive, locked into his supply of money (or promise thereof) until he has bought his product in exchange for money, and consumption of the good(s) then takes place over time. A utility value therefore exists with the consumer immediately *before* the point of purchase, is confirmed on purchase and then declines as consumption begins to take place. Entropy, on the other hand, is a property that is released from a product as it goes through a process of consumption. For example, if fuel is burnt, useful exergy value is consumed and lost forever, and entropy is created and dispersed as a Second Law loss. The difference therefore is that utility resides with the *purchaser or 'owner'*, and entropy resides with the *purchased* or *'product'* (being released to the ecosystem). Utility therefore might be defined as *potential economic entropic value* to be released on consumption.

A further point to consider is that utility decisions occur continuously, as output value flow proceeds. Thus changes in both utility and entropic value in an economic context occur over time. In the case of entropy, this is through the index of trading value **T**.

As a first requirement, any thermodynamic process considered must be able to provide the impetus to allow price and utility to vary quite independently of the underlying productive content. From the First Law of Thermodynamics, the work done **W** alone will not provide such an impetus, since it is concerned *only* with changes in volume flow. The most likely candidate therefore is the entropic value **Q** added or taken away, since that constitutes essentially new money, abundance and scarcity and changes in demand, all of which can influence consumers' purchasing preferences.

Reverting to our thermodynamic development, the polytropic case $PV^n = Z$ represents a suitable process to examine as, from figure 3.12, all of the other processes can be derived from this.

Considering a *single* good, then it will be recalled that the equation for a reversible polytropic economic process, set out at equation (3.51), is equal to:

$$Q = Nk\left(\omega + \frac{1}{1-n}\right)(T_2 - T_1) \tag{3.69}$$

Where **Q** is the Entropic value being added or taken away during the process. In differential terms this can be restated as:

$$dQ = Nk\left(\omega + \frac{1}{1-n}\right)dT \qquad (3.70)$$

By dividing both sides by incremental work done for a reversible process **dW =PdV**, we could re-state this as:

$$\frac{dQ}{PdV} = Nk\left(\omega + \frac{1}{1-n}\right)\left(\frac{dT}{PdV}\right)$$

Substituting in the ideal economic equation **PV=NkT**, we have:

$$\frac{dQ}{PdV} = \left(\omega + \frac{1}{1-n}\right)\left(\frac{dT}{T}\right)\left(\frac{V}{dV}\right) \qquad (3.71)$$

And from equation (3.55) for the polytropic process we have in differential form:

$$\frac{dT}{T} = (1-n)\frac{dV}{V}$$

Substituting this into equation (3.71) and re-arranging we have:

$$\frac{dQ/dV}{P} = (1-n)\left(\omega + \frac{1}{1-n}\right) \qquad (3.72)$$

or

$$\left(\frac{dQ/dV}{P}\right) = (\omega - \omega n + 1) \qquad (3.73)$$

And in the alternative we can write:

$$\left(\frac{dQ}{dW}\right) = (\omega - \omega n + 1) \qquad (3.74)$$

Equations (3.73) and (3.74) say:

- For one particular good, incremental change in Entropic Value with respect to incremental volume change **dQ/dV**, all divided by price **P** is equal to a simple function of the value capacity coefficient ω and the elastic index **n**, and in the alternative:

- Incremental change in Entropic Value with respect to work done **dQ/dW** is equal to the same function of the value capacity coefficient ω and the elastic index **n**.

These equations have some significance as, when compared to equation (3.68) for the Law of Diminishing Marginal Utility, they imply that utility for a single good might be related to entropic value **Q**, and that marginal utility for a single good might be related to incremental change in entropic value with respect to volume change **dQ/dV**, which is also a function of the index of trading value **T** multiplied by the incremental change in entropy with respect to volume change i.e. **dQ/dV=TdS/dV**.

This is *not* the same, however, as saying that the marginal utility of a single good is purely a function of entropy change. The index of trading value **T** comes into play as well because utility values occur in the context of streams of value flow **PV** over time. Moreover, from equation (3.69) it can be seen that values for the right hand side are dependent upon the type of process considered. For the five processes set out in this chapter, the right hand side of equation (3.73) is defined as follows:

Process	*Elastic Index*	*Marginal Entropic Ratio* $(\omega - \omega n + 1)$
Constant volume	n=∞	-∞ (no volume change)
Constant price	n=0	(ω + 1)
Iso-trading	n=1	1
Polytropic	n=n	(ω - ωn + 1)
Isentropic	n=γ	No value as entropy change is zero

Thus comparisons of entropy value changes for processes with different elastic indices **n** and value capacity coefficients ω will yield different answers. We will call the factor (ω - ωn + 1) the *Marginal Entropic Ratio* for a single good, being equal to marginal entropic value with respect to work done **dQ/dW**; equal also to **TdS/PdV**. Figure 3.16 sets out a graphical representation of the Marginal Entropic Ratio to explain the differences.

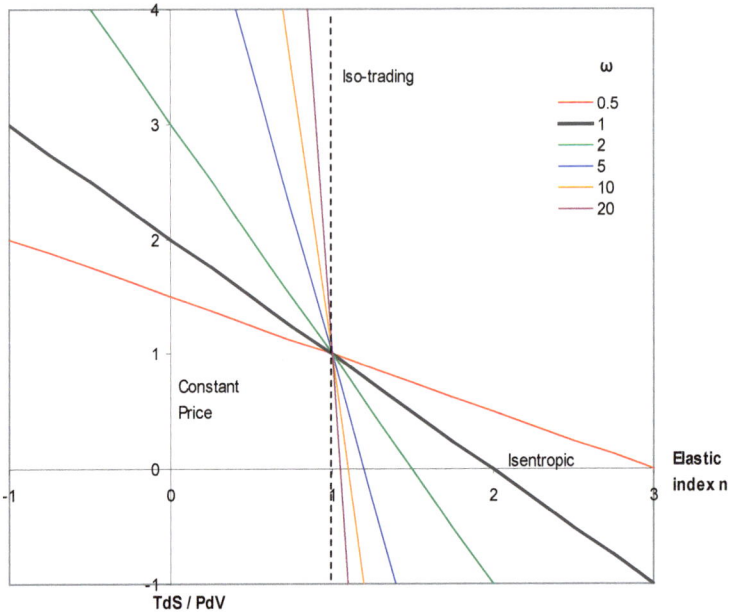

Figure 3.16 Marginal Entropic Ratio (ω – ωn + 1) versus Elastic Index n

Finally, if we substitute **TdS** for **dQ** in equation (3.73), we have:

$$\left(\frac{TdS/dV}{P}\right) = \left(\omega - \omega n + 1\right)$$

Thence:

$$\left(\frac{dS/dV}{P}\right) = \left(\frac{1}{T}\right)\left(\omega - \omega n + 1\right) \tag{3.75}$$

The same result is obtained if we consider a unit stock process, since entropy **S** and volume flow **V** are replace by **s=S/N** and **v=V/N**. Thus:

$$\left(\frac{ds/dv}{P}\right) = \left(\frac{1}{T}\right)\left(\omega - \omega n + 1\right) \tag{3.76}$$

All of the forgoing analysis suggests that there is likely to be a strong relationship between the economic concept of utility Y and the thermodynamic concepts of entropic value Q and entropy S. The relationship also depends however upon the value capacity coefficient ω, (encompassing the lifetime and aesthetic values attached by humans to a product), and the elastic index n, both of which may vary between goods.

More work is therefore required to examine the relationship, and in particular, the situation of multiple goods and partial utilities within a budget constraint.

CHAPTER 4   PRODUCTION AND ENTROPY PROCESSES

The traditional economic theory of production expounded by Solow [Solow, R. M. (1956, 1957)] is built on the theory of the firm, in terms of production functions and marginal productivity equations as a consequence of profit maximisation. The production function is regarded as a technical relation between inputs and outputs. It is assumed that output V_O per unit of time of goods and services is a function of capital stock N_K *[constant price money terms]* and labour stock N_L, to which is added an exogenous technical progress function to determine long term demand. Land can also be added, along with a feedback mechanism of investment from profit surpluses to augment capital stock. Equation (4.1) sets out a simple form of the well-known Cobb-Douglas production function [Cobb, Douglas (1928)] with a Hicks-neutral technical progress function added [Hicks, J. (1932)] – viz:

$$V_O = \left[Ae^{\lambda t}\right]\left(N_K\right)^\alpha \left(N_L\right)^\beta \text{ or } \left(\frac{V}{V_*}\right)_O = e^{\lambda(t-t_*)}\left(\frac{N}{N_*}\right)^\alpha_K \left(\frac{N}{N_*}\right)^\beta_L \quad (4.1)$$

Where A is a constant incorporating an output usage rate per unit of time per unit of stock **N**, * is a starting position, and the factors α and β are elasticity coefficients, signifying:

$\alpha + \beta < 1$ *Decreasing returns to scale*

$\alpha + \beta = 1$ *Constant returns to scale*

$\alpha + \beta > 1$ *Increasing returns to scale*

For an entire country, output volume flow would be measured as final output value (GDP) modified by a price index to convert it to a volume basis. Likewise capital stock would be measured in constant price terms.

The equation is forward looking, effectively projecting exponential growth into the future and, with no consideration of constraints that may halt the process, does not take account of short-term business cycle variations. The economist tends to take for granted that resources are used up in the process. As stated earlier in this book, resource values are deemed 'free', and consideration is given only to capital stock and labour costs. In the ultimate if there were no resources, one could still project forward economic growth on the basis of existing capital stock and labour with a technical progress function.

The physicist therefore would regard the economic approach to production as contravening the Laws of Thermodynamics. Heat can only flow from a hotter body to a cooler body, and the factors providing the heat and energy input (technically exergy in relation to an environment, or other measure of productive content) are mostly the resources, with some labour and capital stock consumption, which in turn create product, net of an efficiency factor, with an associated release of entropy to the environment. The proverbial ant heap does not grow of its own, but by the consumption of accessible resources and nutrients. This is the theory that is championed by Schrödinger, Schneider and Kay, of living organisms maintaining a local level of organisation at the expense of producing entropy in the environment.

Ayres et al have shown that traditional economic theory explains very little of growth, compared to factors such as natural resource exergy, with the most important factor of technical progress driving output being related to the improvements in efficiency with which fuels are converted into useful forms of work.

Proceeding further, the consumption of output from the production process constitutes the means by which humankind is able to regenerate itself over time, and it follows therefore that this process of regeneration can also be considered to be a production process, though having a much longer lead time because of the cycle of births, education and upbringing to maturity. It is a continuation of the production process to the next stage, with residual output outside of this process finding its way into accumulated consumption. This perspective is somewhat the reverse concept to economics, which tends to regard consumption as the raison d'être, with only a part of consumption finding its way to re-births and education.

4.1   A Simple Production System

We imagine a simple production system involving a fixed process. By this is meant that to produce a product, specific amounts of the productive content of particular capital stocks, labour and resources/materials are required, and no other combination, and which are brought together under a specified production process. Substitutes are therefore excluded. As output is created, some waste through efficiency losses is rejected. Some of this waste is recovered via subsidiary processing production units.

Capital Stock

Labour Stock

PRODUCTION
PROCESS

Output Stock

Resource Stock

Waste

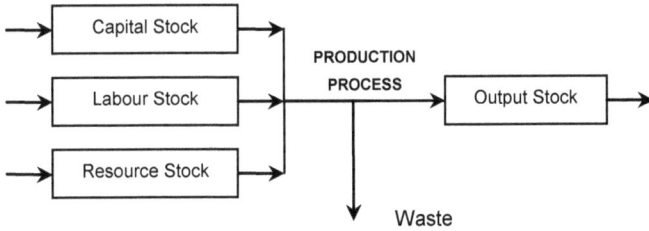

Figure 4.1 A Simple Production System

In gas/chemical reactions such fixed arrangements are normal, there being only one configuration of inputs and outputs; for example, two molecules of carbon monoxide combust with one molecule of oxygen to form two molecules of carbon dioxide. There is no other combination. Thus in an economic system with an inexact mixture, if a reaction proceeds so that one input is completely used up, there are likely to be some other inputs left over. If there is a scarcity such that a substitute is found, then the substitute is another product made by another process/system, and outside the dynamics of the system under analysis. Likewise, if a different mix of inputs can be engineered by investment in plant and management, we are again discussing another system. In this initial analysis it is only the dynamics of a fixed system that are being examined, and not feedback mechanisms to change the choice or structure of a system.

There is therefore perhaps a difference between human economic and gas/chemical systems, in that humankind can continually modify its economic system in order to change the objectives and benefits – a complex feedback mechanism – though the section on entropy maximisation later in this chapter might give cause to be cautious regarding making such an assertion. A further factor to consider is that the inputs and outputs in human economic systems are all very different in nature, though this does not stop the process of combining them together. A part of this of course occurs through the medium of money, which travels in the opposite direction to product inputs and outputs.

Imagine a specific production process, whereby to form **x** units of a particular output product **O** over time, requires consumption of **a** units of capital plant **K**, **b** units of labour **L** and **c** units of resource(s) **R**. **y** units of residue waste **D** are also produced *[excluding that accounted for in the GDP i.e. we are considering efficiency losses and waste ejected to the ecosystem/environment]*. The factors **a**, **b**, **c**, **x**, **y** are set by the nature of the productive content of the specific process considered: hours of particular human expertise, energy and electricity requirements, nuts & bolts from a resource stock. They indicate the relative amounts that are required to make up a product or service. We could write:

$$aK + bL + cR \Leftrightarrow xO + yD \qquad (4.2)$$

Where the double arrow signifies that the process is not necessarily a complete one; it depends upon the actual relative concentrations and constraints posed by each component to promote the forward path. We could simplify the above equation by dividing through by x, and substitute $\alpha=a/x$, $\beta=b/x$, $\delta=c/x$ and $\rho=y/x$. Thus:

$$\alpha K + \beta L + \delta R - \rho D \Leftrightarrow O \qquad (4.3)$$

As set out at the beginning of this book, the thermo-economic approach is based on the Le Chatelier Principle, which states: *"If a change occurs in one of the factors under which a system is equilibrium, then the system will tend to adjust itself so as to annul as far as possible the effects of that change".* This principle was espoused also by the economist Samuelson (1964), and is one that the author has used in previous papers (1982, 2007 & 2008). It is not asserted that a system *will* attain equilibrium, only that it will continually seek to proceed to such a position. In fact such a system will not work without being in a state of disequilibrium.

Straight away we meet with a stumbling block, in that in an economic system resources are not defined in the manner which a physicist would understand. The value of resource consumption entering an economic system is defined by reference to the wage and profit attached to it, not the productive content. At a macro-economic level no resources enter the system except those entering via import expenditures, but set off by export sales. It is of course recognised that added value and final output can be divided up into specific industrial and government sectors, but in money terms the value of product flow *[net of efficiency losses]* is balanced out exactly by the consumption costs of labour wage and capital stock in this. *[We are accepting that this is a somewhat simplistic view of an economic system, ignoring items such as taxation.]* Macro economic accounts automatically assume that the position is balanced. So where is the motive force which drives the system forward?

A possible solution to this is to redefine input and output flows. Suppose we separate out for each component the volume flow part that is *active* $\mathbf{V_a}=v\mathbf{N_a}$

[the specific volume rate multiplied by the active part of a stock N_a], and the volume flow part that is *inactive* $V_c = vN_c$, that is, *not* currently contributing to the production process flow *[by the inactive part of the stock N_c]*, as in the diagram at figure 4.2. An example would be the actual output flow produced by *active* employed labour, and offset on the other side, the *potential* output flow that could otherwise be generated by the remaining *inactive* labour that is currently unemployed. If there is a demand for increased output, then there are additional units of labour available to fill the productive gap, and vice-versa.

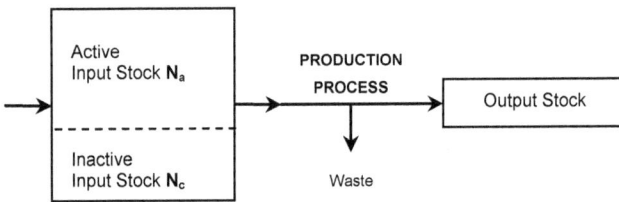

Figure 4.2 Active – Inactive Components of Economic Flow.

We could repeat this for all of the components of the flow. For a producer capital stock it is reasonable to posit that the generation of profit is more likely to be enhanced if all of the capital stock was utilised at all times and with no 'down-time' when it is unused. In the case of a money stock, servicing the output and consumption flows of a whole economy, a reduced or enhanced level of utilisation could be engendered by changes in interest rates.

In the case of a resource stock, although at a macro level it does not appear in national accounts, except as a value flow attributed to labour and profit or added value output after efficiency losses *[divided up by industry sector]*, it does nevertheless exhibit properties of utilisation, in that the extent to which the production flow rate can increase or decrease will be influenced by demand and by the levels of depletion and discoveries. In reverse to a resource stock also, one could imagine a waste stock, the size or impact of which might influence the level of output flow from an economic process, via pollution or other effect from the environment and the ecosystem. Finally, we can imagine a flow of product to an output stock might occasion an increase in saturation or *inactivity* of that stock, unless it is depleted by factors of depreciation, consumption and obsolescence.

Now we consider some *arbitrary* reactant factors of capital plant, labour and resource usage, depicted by the initial dotted lines in figure 4.3, which

93

feed output production at initial steady flow rates of defined *activity* and *inactivity* levels, that is, the proportions of each of the inputs that are either actively or not actively contributing to output production, these proportions not necessarily being the same for each.

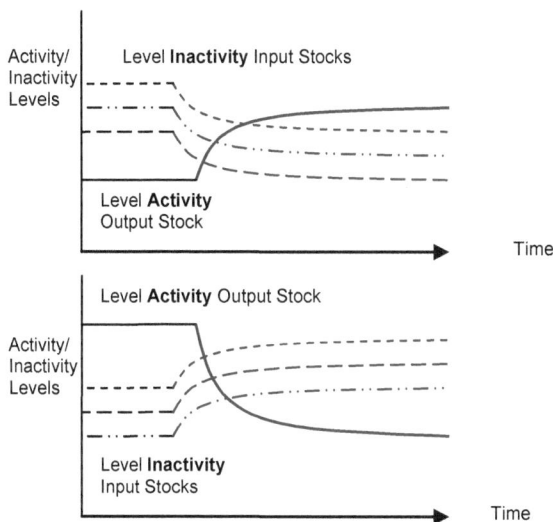

Figure 4.3 Activity/Inactivity Levels in a Simple Production Flow Process

Imagine that the system is then provided with an additional *motive force or utility value* potentially to encourage the flow rate of output, such that the *activity* level of output can rise, as in the top chart. Such a rise can be met by increasing the *activity* levels of the input stocks and reducing their *inactivity* levels. What levels they each reach, however, will depend *both* upon the ultimate demand for output flow, *and* the availabilities of each of the input flows. We could however equally consider a position where the *motive force* or *utility value* might be such as to reduce rather than increase the flow rate of output; in which case the *activity* levels of the inputs would fall and their *inactivity* levels would rise, as in the lower chart.

It is plain that the ultimate level reached will be fixed by which of the input and output flows that constitute the major limiting factor. An input factor that was already fully active could not support a rise in output and would become the dominant restraining force at that moment in time. It is not axiomatic that input and output flows will rise in tandem with each other.

Potential restrictions might be particularly prevalent in one or more of the factors, compared to the others, and impact at a different time to the others, as the economic system proceeds. For example, the availability of labour might be restricted; a particular material might be in short supply; a restriction might exist with regard to the creation of addition waste, such as additional CO_2; or the level of financing costs to fund an increase in demand and output flow, in terms of interest payments, might subtract from fully increasing production. Figure 4.4 illustrates the principle, and the movement to a new equilibrium position, occasioned by the demand motive/utility value, which could be either positive or negative, depending upon the situation at a point in time.

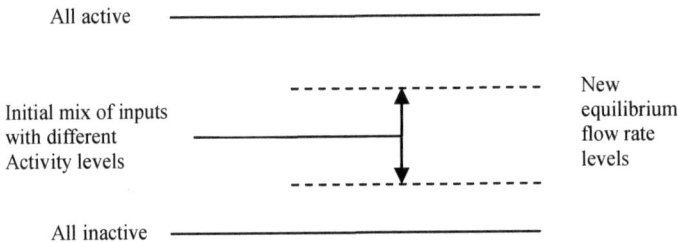

Figure 4.4 Active / Inactive Flow & Equilibrium Level

In order to represent the potential work value occasioned by the impact of a motive force/utility value initiating increased or decreased activity flow rates of capital stock, labour and resources and converted into a changed product flow rate, resort can be made to two thermodynamic properties known as the Helmholtz Free Energy function **F** *[German physicist Hermann von Helmholtz 1821 – 1894]* and the Gibb Free Energy function **X** *[American physicist Josiah Willard Gibb 1839 – 1903; we have used the symbol X rather G, as G in this book is taken to be value flow G].* These functions are common concepts in the thermodynamic analysis of chemical and gas reactions, particularly for closed systems, though there is no reason why they should not be used also for flows of inputs and output per unit of time, in which case we are considering Free Energy per unit of time. They express the total amount of *Exergy* (available energy) which can be used up or passed on during a reaction to equilibrium. In thermodynamic terms they respectively have the formulae:

$$dF = -(PdV + SdT) \text{ and } dX = (VdP - SdT) \qquad (4.4)$$

Where **P** is pressure, **V** is volume, **T** is temperature and **S** is entropy.

In economic terms the free energy might better be described as Free Value, being the amount of additional or reduced useful *value* available per unit of time that can be used up during a reaction between inputs such as labour, capital stock and resources, potentially to produce additional or reduced output product flow. Equations (4.4) could therefore represent also an economic equivalent, relating Free Value to price/cost **P**, volume flow **V**, entropy **S** and the index of trading value **T**.

The choice of which function to use is a matter of preference. The Helmholtz function works in terms of partial volumes, whereas the Gibb function works in terms of partial pressures *[economic equivalent price]*. The choice made here is the Helmholtz function **F** though whichever function is used the result in economic terms is essentially the same.

Continuing with the Helmholtz function, the assumption is made that immediately before the actual point of increased/decreased conversion rate of inputs into outputs, no change in the index of trading value **T** of either inputs or outputs has occurred. We can see what this means from equation (2.21), chapter 2:

$$PV = PvN = NkT \quad \text{or} \quad Pv = kT = \text{Constant} \qquad (4.5)$$

Thus value flow rate **Pv** per unit of stock, equal to price **P** multiplied by specific volume rate **v,** is assumed to be constant for both inputs and outputs, irrespective of the number of units in the stock. The productive content **k** is of course constant. This equates to the *iso-trading* process met in section 3.5, chapter 3. The factor **v** for a unit of stock has some similarities to the chemical concept of an 'activity coefficient' applying to the total concentration of stock **N** available, to equal the effective net input or output in the reaction. Thus **V=vN** as in equation (4.5). It will be recalled also from equation (2.2) that the factor **v** was equal to:

$$v = \left(\frac{1}{\xi_t} \right)$$

Where ξ was the ratio of the natural lifetime t_L of a unit in a stock, compared to the standard transaction time t_t (usually a year). This differs from stock to stock. For example, electronic money has a relatively short lifetime, arable food cycles takes place over a year, and human labour cycles over 40-50 years. Further, the cycles of each individual item of the same type can be at different stages of their lifetime. For example, young people exist alongside older people, and some items of capital stock can be new, compared to others that have reached the end if their useful life.

Returning to our development, because we have assumed $dT = 0$, then for either input or output flows at the system boundary, equation (4.4) can be reduced to:

$$dF = -PdV \qquad (4.6)$$

Which is the negative of the incremental work done dW that we encountered in chapter 3. Hence, for a spontaneous reaction to take place to produce additional output flow, consumption (reduction) of free value F occurs, that is dF is negative. By substituting in $PV=NkT$ we have:

$$dF = -NkT\left(\frac{dV}{V}\right) \qquad (4.7)$$

It will be noted that this expression is similar in construction to the Iso-trading process, referred to earlier, and where the entropy change at section 3.5, chapter 3 was stated as:

$$dS = Nk\left(\frac{dV}{V}\right)$$

Thus a change in free value equates to an opposite change in entropy, adjusted by the index of trading value T.

$$dF = -TdS \qquad (4.8)$$

Readers will note also that this equation is identical in form to incremental entropic value change $dQ = TdS$, encountered at section 3.9 of chapter 3, and therefore that free value F is related to the negative of the economic concept of utility Y – see equations (3.73) and (3.75). So now we have a *motive force/utility value* that connects directly to the *free value* production function.

By integrating equation (4.7), the free value F inclusive of that for the equilibrium flow volume level of the *active* output flow and for each of the *inactive* potential input flows, can be stated as:

$$F = F_* - NkT \ln(V) \tag{4.9}$$

Where the suffix * denotes an equilibrium position.

Now although the flow rates of the input and output reactants are initially at *arbitrary* levels, according to equation (4.3) however, the reactants have to combine/react in fixed proportions. Thus we combine all the free values of the *active* and *inactive* components available in accordance with the fixed proportions, where the suffixes **a** equals the *active* component and **c** the *inactive* component:

$$\alpha[F_c]_K + \beta[F_c]_L + \delta[F_c]_R \Leftrightarrow [F_a]_O + etc \tag{4.10}$$

Where

$$[F_a]_O = \alpha[F_a]_K + \beta[F_a]_L + \delta[F_a]_R + etc \tag{4.11}$$

Thus in equations (4.10) and (4.11) free values of *inactive* input components effectively become *active* components and join together to make *active* output.

Now substituting in from equation (4.9) we have:

Free value of inputs *[the inactive components]*:

$$= \alpha[F_{*c} - NkT \ln(V_c)]_K + \beta[F_{*c} - NkT \ln(V_c)]_L + etc$$

Free value of output *[the new active components]*:

$$= \alpha[F_{*a} - NkT \ln(V_a)]_K + \beta[F_{*a} - NkT \ln(V_a)]_L + etc$$

Where F_{*a} and F_{*c} represent the free values at the equilibrium state between the active and inactive flow positions. Thence the change in free value accompanying the reaction to the equilibrium state is the difference between these two:

$$\begin{aligned} \Delta F &= \alpha[F_{*a} - NkT \ln(V_a)]_K + etc \\ &\quad - \alpha[F_{*c} - NkT \ln(V_c)]_K - etc \end{aligned} \tag{4.12}$$

At this stage the values of **N, k** and **T** of the inputs and outputs all have very different forms, flow rates and efficiency losses. We are comparing apples with pears, humans with capital stock and resources, each with very different lifetimes and productive content and prices *[The same occurs when economists compare utilities]*. However, we do know that in a production process they combine together to produce output, and that a complementary reverse flow of money occurs when they do this. We could therefore replace the value (**NkT**) for each factor by a monetary equivalent. Further, because this part of each reactant is now defined in terms of an exchangeable medium, we could remove this right outside of the square brackets and assume a global money function for this, and while the weights of each factor will then be different, we could continue with the development on the assumption that appropriate adjustments to factors **α, β** etc can be made to compensate, and incorporating also any relative changes in efficiency losses. Thus after rearranging equation (4.12) we have:

$$\Delta F = \Delta F_* - \left(NkT\right)_{money} \ln\left[\frac{\left(V_a\right)_K^\alpha \left(V_a\right)_L^\beta \left(V_a\right)_R^\delta \cdots}{\left(V_c\right)_K^\alpha \left(V_c\right)_L^\beta \left(V_c\right)_R^\delta \cdots}\right] \qquad (4.13)$$

Where **ΔF*** is the change in free values of inputs and outputs accompanying the reaction, when all the reactants and products are at their new equilibrium flow rates.

Now when the reaction has reached equilibrium, the change in free value **ΔF** becomes zero, consequently from equation (4.13) we can write:

$$\Delta F_* = \left(NkT\right)_{money} \ln\left[\frac{\left(V_a\right)_K^\alpha \left(V_a\right)_L^\beta \left(V_a\right)_R^\delta \cdots}{\left(V_c\right)_K^\alpha \left(V_c\right)_L^\beta \left(V_c\right)_R^\delta \cdots}\right]_* \qquad (4.14)$$

Where the subscript * denotes the mix for the system at equilibrium. Since the equilibrium free value change **ΔF*** is the defined state of unit activity of the mix, it is apparent that **ΔF*** must be constant, and it follows that the part of equation (4.14) contained in the square brackets must be constant too. Thence we can write:

$$\left[\frac{\left(V_a\right)_K^\alpha \left(V_a\right)_L^\beta \left(V_a\right)_R^\delta \cdots}{\left(V_c\right)_K^\alpha \left(V_c\right)_L^\beta \left(V_c\right)_R^\delta \cdots}\right]_* = \psi \qquad (4.15)$$

where Ψ may be called the *Equilibrium Constant*, representing the new flow rate position.

As time goes on, however, the equilibrium flow rate position may change, owing to the dynamic interactions and feedback mechanisms of all the factors. For the time being, however, we will continue to regard the system as a fixed one, and then consider the position when this constraint is relaxed.

By substituting equation (4.14) back into equation (4.13) we have an equation for the free value change to the new equilibrium flow rate for the process:

$$\Delta F = -(NkT)_{money} \ln \left[\left[\left(\frac{V_K}{V_{K*}} \right)^\alpha \left(\frac{V_L}{V_{L*}} \right)^\beta \left(\frac{V_R}{V_{R*}} \right)^\delta \cdots \right]_a \left[\left(\frac{V_{K*}}{V_K} \right)^\alpha \left(\frac{V_{L*}}{V_L} \right)^\beta \left(\frac{V_{R*}}{V_R} \right)^\delta \cdots \right]_c \right]$$

(4.16)

By turning round equation (4.16), and substituting unit money free value change Δf for $\Delta F/N$ we can write:

$$\left[\left(\frac{V_K}{V_{K*}} \right)^\alpha \left(\frac{V_L}{V_{L*}} \right)^\beta \left(\frac{V_R}{V_{R*}} \right)^\delta \cdots \right]_a = \left[e^{\left(\frac{-\Delta f}{kT} \right)_{money}} \right] \left[\left(\frac{V_K}{V_{K*}} \right)^\alpha \left(\frac{V_L}{V_{L*}} \right)^\beta \left(\frac{V_R}{V_{R*}} \right)^\delta \cdots \right]_c$$

(4.17)

Thus a change in the *active* components of output volume flow to their new equilibrium levels is equated to a function of the change in the *inactive* components of output flow to their new equilibrium levels, all multiplied by an exponential money free value factor at the front of the equation.

A simplified means of representing equations (4.17) is to take out all the equilibrium amounts, since their product for the equilibrium flow rate is a constant, as per equation (4.15), and put them into the constant ψ. Thus:

$$\left[[V_K]^\alpha [V_L]^\beta [V_R]^\delta \cdots \right]_a = \left[\Psi e^{\left(\frac{-\Delta f}{kT} \right)_{money}} \right] \left[[V_K]^\alpha [V_L]^\beta [V_R]^\delta \cdots \right]_c \qquad (4.18)$$

Or, since output volume flow is a function of all the *active* components, we can write in shorthand:

$$[V_O]_a = \left[\Psi e^{\left(\frac{-\Delta f}{kT}\right)_{money}} \right] \left[[V_K]^\alpha [V_L]^\beta [V_R]^\delta \dots \right]_c \qquad (4.19)$$

Or:

$$[V_O]_a = \left[\Psi e^{\left(\frac{-\Delta f}{kT}\right)_{money}} \right] [V_O]_c \qquad (4.20)$$

Equating the *active* contribution of output flow to the *inactive* parts of the inputs.

Equation (4.19) has some similarities to that defined in the Arrhenius Equation *(Swedish scientist Svante Arrhenius 1859–1927)* and the Eyring Equation *(American scientist Henry Eyring 1901–1981)*, which are mathematical expressions used in chemical kinetics to link the rate of reaction to the concentration of the reactants. The similarity is particularly striking if the volume rates are split into their specific volume rate v and stock number N, thus $V=vN$:

$$[V_O]_a = \left[\Psi e^{\left(\frac{-\Delta f}{kT}\right)_{money}} \right] \left[[N_K]^\alpha [N_L]^\beta [N_R]^\delta \dots [v_K]^\alpha [v_L]^\beta [v_R]^\delta \dots \right]_c$$

Or:

$$[V_O]_a = v_Z \left[\Psi e^{\left(\frac{-\Delta f}{kT}\right)_{money}} \right] \left[[N_K]^\alpha [N_L]^\beta [N_R]^\delta \right]_c \qquad (4.21)$$

Where v_Z is some composite specific volume rate relating to the product formed.

Thus in an economic system, change in the *active* output flow rate is a function of the *inactive* components available. A restriction in any particular reactant can affect the forward rate, and likewise, abundance of the same reactant will place weight on the other reactants to be potential constraints on output flow. For example high availability of *inactive* resources such as energy and food resources will encourage growth in labour (population) and capital plant.

A number of further points can be made.

First, it will be recalled from equation (4.8) that incremental free value **dF** is equated to the negative of incremental entropic value **TdS** *[and therefore utility – see section 3.9]*. This will apply as much to the 'money' part of the equation as to all the other components. i.e.:

$$dF = -TdS \text{, and for unit money stock: } df = -Tds$$

Substituting this effect into equation (4.20) we could write:

$$\left[V_O\right]_a = \left[\Psi e^{\left(\frac{\Delta s}{k}\right)_{money}}\right]\left[V_O\right]_c \qquad (4.22)$$

And by turning round equation (4.22), taking logs and differentiating (remembering that the equilibrium factor ψ has been defined as constant), we can write:

$$ds = k_{money}\left[\frac{dV_{Oa}}{V_{Oa}} - \frac{dV_{Oc}}{V_{Oc}}\right] \qquad (4.23)$$

Where **k_money** = 1 (£1, $1 etc). Writing out all the inactive components in full we can put:

$$ds = k_{money}\left[\frac{dV_{Oa}}{V_{Oa}} - \alpha\frac{dV_{Kc}}{V_{Kc}} - \beta\frac{dV_{Lc}}{V_{Lc}} - \delta\frac{dV_{Rc}}{V_{Rc}}....\right] \qquad (4.24)$$

which indicates that the rate of change in *active* output flow **dV_Oa/V_Oa** is dependent in part upon the *motive force/utility value* introduced by the incremental entropy change **ds**. If this is positive, output value flow could rise and if it is negative output flow could go down, but depending also upon the rates of flow change of the other factors. Thus a positive entropy change indicates a motivating force to promote the forward path. Comparison of this equation with equation (4.1) highlights the difference between the thermodynamic and the traditional economic approaches.

Second, it will be noted that the incremental entropy change **ds** effectively has the property of percent rate of change, as with rates of interest, factor growth and inflation, particularly as **k_money** is commonly defined as **1**, for any currency.

Third, it should be noted that in equations (4.23) and (4.24) we are dealing with a *particular* production process, with flow of output production and flow of input consumption on opposite sides of the process. We are *not* allowing substitution of one reactant by another external reactant to effectively change the system.

The dynamic nature of the analysis becomes apparent when the process is joined up with all the other adjacent interacting systems, and likewise they in turn are linked up with more distant systems. Thus capital stock depreciates and is replaced by new investment; labour retires and is replaced by recruitment of new personnel, and resource stocks are replenished by reordering. If one reactant is used up - for example, product **O** is sold on to the next system - then the system produces more **O**. Likewise as a resource **R** is used up, it is replaced from another system, which perceives demand for its product, and so the reaction goes on. We have derived an inter-reacting trading process, with free value continually being used up by one system and replaced by that of another, and economic entropy change arising. Thus if a particular constraint has an effect on a connecting system or one further up or down the line, then a chain reaction might occur.

We can see that such a group of systems could then approach that of an economy, as illustrated by figure 2.6 in chapter 2. Ultimately, from an economist's viewpoint, output at the macro level, through an input-output process, also forms part of input, and their summations are equivalent – the circular notion of economic flow. The difference from the thermo-economic viewpoint is that this circular flow is actually fed by nature's abundance.

Proceeding further, if the equilibrium constant ψ was such that no constraint was of particular significance, as input reactants are replenished and output is removed, then in such a 'golden age' the rate of change of output flow would be dependent only upon the incremental entropy change **ds**.

However, accepting that particular constraints do occur in life and not necessarily all at the same time *[such as full employment, high interest rates, commodity shortages etc]*, then the equilibrium position will likely not be constant at all times as initially assumed. Indeed, it may be changing as different constraints act upon an economic system at different times. Figure 4.5 shows the effect of movements between outputs flow and constraints and the entropy change or motivating force / utility value.

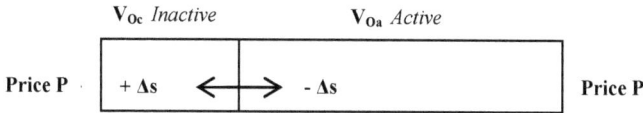

Figure 4.5 Active and Inactive Output flow, and entropy change

Finally, it may be noted that provided *both* the *active* and *inactive* parts on each side of equations (4.22) and (4.23) can be shown to have the same rate of change in price **P**, then we could replace volume flow **V** with value flow **G** (=**PV**), and obtain the same representation. We are not allowing substitution where the price of a factor can be bid up or down against a substitute or a different mix of input and output factors – this is a fixed system. Therefore, making this substitution we could write:

$$ds = k_{money} \left[\frac{dG_{Oa}}{G_{Oa}} - \frac{dG_{Oc}}{G_{Oc}} \right] \qquad [subject\ to\ \mathbf{dP/P}\ active = \mathbf{dP/P}\ inactive]$$

(4.25)

To repeat, this adjustment will *only* be applicable to situations where the rate of change in price is the same for both *active* and *inactive* parts. An example of this is an interest rate affecting the impact of the *active* and *inactive* parts of money. One has either to subtract the inflation rate from the gross interest rate to compute a 'real' *net* interest rate and its effect on output volume flow, or to multiply price by output volume flow to equal output value flow **G = PV**, and compare rates of change in **G** with the *gross* interest rate.

The beauty of equations (4.22) – (4.25) is that they describe the motivating force of an economic system, which will be amplified in successive chapters of this book, concerning money, employment, interest rates and energy. **ds=0** represents a position where rate of change in *inactive* flow loss matches that of output, and **ds<0** represents a position where *inactive* flow loss is growing faster than that of output. In such a position eventually output growth is forced to slow down or decline, depending upon the structural relationship between *active* and *inactive* components. Likewise when **ds>0**, and *inactive* components are receding, then growth in output value flow can resume or increase faster.

From the analysis developed it can be seen that the key determinants of the rate of economic flow are:

- The levels of disequilibrium free value change Δf and entropy change Δs. The more negative the former and the more positive the latter, the higher the forward rate of change of reaction.

- The index of trading value T. The higher this, the higher the level of output value flow per unit of time.

- The relative level of resource availability, compared to capital and labour stocks. The higher the availability of resources, the more capital and labour can be utilised to produce output, with a consequent feedback to create more capital and labour stocks. Likewise a restriction in resource availability will slow down or reduce the forward path.

- Similarly, the lower the level of accumulated waste (or the greater efficiency with which it is removed and returned and absorbed into the environmental and ecological cycles) the higher the potential output level.

- The level of consumer stocks – the lower these are the higher the potential output level. Thus high obsolescence, with low lifecycle and low saturation promote the forward path, though this does little to conserve the resources and energy from which the stocks are made.

- The effect of any particular constraints.

The nature of the thermodynamic exposition indicates that output is a complex, dynamic function of stocks and flows, with elements of systems dynamics, with an equilibrium position that is continually varying, moving to the left or to the right, depending upon the stock quantities and any constraining factors. It is the nature and position of the constraining inactive factors that determines output growth or decline. Exponential growth of output is not axiomatic.

4.2   Reaction Kinetics

The analysis set out so far gives rise to a number of basic kinetic models and principles relevant to a thermodynamic theory of economics, which will now be described. First, if we suppose that the number of stocks is reduced to a single economic component, with a stock of known absolute size N that is being used up, and which cannot be replaced or augmented by successful prospecting, and where the lifetime of a stock item is a constant $t_L = 1/\varphi$, then the specific volume rate v is equal to φ, and we could define the volume flow rate V as:

$$V = \frac{dN}{dt} = -\varphi N \tag{4.26}$$

The mathematics of this reaction is simple, being a declining exponential function, with the formula representing a declining balance stock with a depreciation or depletion rate of $v = \varphi$.

$$N = N_0 e^{-\varphi t} \tag{4.27}$$

Figure 4.6 illustrates the dynamics of a single stock. The stock has a half-life, rather like a radio-active substance.

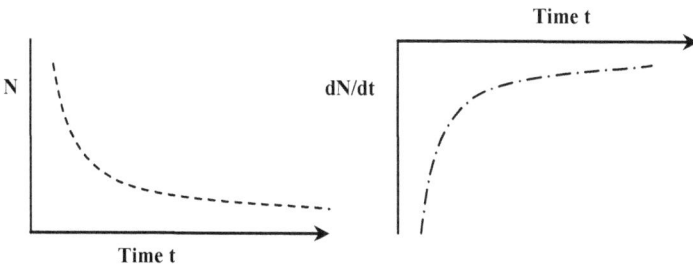

Figure 4.6 Single Declining Stock

One can imagine also, that if a process is proceeding along a constant volume *flow* path, and is then subject to a disturbance of new inputs, or loss of inputs, then there would be an appropriate acceleration or deceleration in the flow, until a new equilibrium flow rate was established.

Capital and labour stocks are of course more complex than the above analysis, having both inputs and outputs, of investment/depreciation, recruitment/retirement and births/deaths via the population cycle. It is nevertheless possible to imagine a two-way process, with a forward reaction having a depreciation rate of φ per unit of time, and a reverse reaction having a creation rate of φ'. Thus V forward equals φN and V reverse equals $-\varphi'N$. Hence net growth in the stock is given by:

$$dN\!\!\Big/\!\!dt = -(\varphi - \varphi')N \quad \text{and} \quad N(t) = N_0 e^{-(\varphi-\varphi')t} \qquad (4.28)$$

Although there are large time lags between inputs and outputs, if the rate of investment φ' is greater that the rate of depreciation φ, then there will be a net growth in the stock, with a growth rate of (φ'-φ). For a two-way reaction of this kind, the equilibrium constant ψ from equation (4.22) becomes equal to:

$$\psi = \frac{\varphi}{\varphi'} \qquad (4.29)$$

The birth rate or rate of investment is of course dependent upon human decision, having regard to current conditions, expectations of the future, but more importantly the availability of resources to fuel growth.

A final process to consider before leaving this section is that of the development of a resource, or a restriction imposed on accumulated output such as consumer stock saturation or accumulated pollution.

When considering a resource with a ready infrastructure to utilise and convert it into product, the likely picture of development is that shown in figure 4.6. The reality of the situation, however, for a resource such as oil or gas, is that the infrastructure required, such as oil-well equipment, power stations, cars, roads, oil tankers, pipelines and all the capital stock needed to use the resource reserves, has to be developed over time. In the early stages therefore output tends at first to grow slowly, being dependent upon the infrastructure in place, before later proceeding at a faster rate. In addition, ultimate knowledge of the extent of resource reserves and their recovery factors are not always known in the development phase of the cycle. This suggests a more complex development involving S-shaped curves of accumulated production and remaining resource; as illustrated in the left hand diagram at figure 4.7, with product accumulation shown by the dotted line and resource reserve remaining shown by the solid line. In the right hand diagram, volume flow accelerates as the curves proceed and then declines as the process draws to its close.

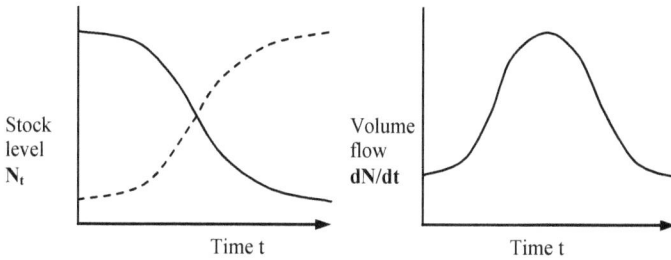

Figure 4.7 Logistic Curves

Such curves belong to the Logistic family of models, of which the Verhulst equation [Pierre Verhulst 1804-1849] and the Gompertz and Sigmoid functions are the more well known. These models find use for example in estimating saturation levels of population and consumer durables. The growth of civilisations tends to follow an S-shaped curve, but some civilisations can also decline, such as the Roman Empire. The Verhulst equation has the following differential form:

$$\frac{dN}{dt} = \varphi N \left(1 - \frac{N}{R} \right) \qquad (4.30)$$

Where **N** is the size of the stock, and **R** is the ultimate size of the stock or resource reserve. The solution to this equation is shown at equation (4.31), where **N₀** is the initial starting level:

$$N(t) = \frac{N_0}{\left[\left(\frac{N_0}{R} \right) \left(1 - e^{-\varphi t} \right) \right] + e^{-\varphi t}} \qquad (4.31)$$

Fitting the above equation to real data is not a simple exercise. Specifically, some knowledge is required of the likely ultimate level of **R,** along with evidence of when the turning point at the middle of the curve at figure 4.7 will occur. All of the above functions are related to differential equations commonly used in systems dynamics. This book is not concerned primarily with the construction of a systems dynamics model, however, though such a model was used by Meadows et al (1972) in 'Limits to Growth', which gained some notoriety in the early 1970s. Nonetheless, at chapter 8 of this book some analyses of world energy, peak oil and climate change are used to try to illustrate the impact of some of the principles of this chapter.

4.3   Entropy and Maximisation

In standard economic utility theory, consumers, faced with a limited income or budget which they can spend, are imagined to choose a particular consumption bundle from a range of possible bundles of goods, services and opportunities at particular prices, such that their perceived utility Y is maximised over time in some manner. Utility decisions of consumers encompass risk [von Neumann-Morganstern 1947], social factors and value [Greene, Baron 2001]. A time element is involved, both in relation to the point at which purchases can be initiated, and the period over which benefits are spread. As consumers proceed through life, each has a developing process that governs their decision-making concerning the opportunities that present themselves, with the nearer-term being more knowable than the longer term.

Corporations are faced with the problem of how to allocate expenditure of a limited budget among a bundle of projects, services and capital expenditures, according to the perceived benefits and risks arising from each, such that their surplus or profit flow will be maximised. Decisions vary from the short-term, such as funding for ongoing output and services, to longer-term ones involving classification of alternative projects and investments. Government agencies also are imagined to maximise the services and benefits that they provide from a limited budget set by tax revenues. As with consumers, the benefits over time are discounted by decreased knowledge of the probability of future outcomes.

The members of all these decision groups have at their cortex innate algorithms which continually guide their choices from among apparent relevant alternatives to endeavour to meet their future needs and desires, as far as they can foretell them. They are genetically inherently programmed in some way to do this. Each may have a different view on this, though one might imagine economic, social, moral, ecological and other factors impacting on this process as to how much their view matches that of their compatriots. To put it shortly, humans are programmed to maximise benefits to themselves and the furtherance of human life, as each perceives their position and qualities.

At chapter 3 a link was established between entropy and utility; and at section 4.1 of this chapter a function was developed linking entropy change to output production and specific constraints. For ease of reference, equations (4.23) and (4.25) are set out again as follows:

$$ds = k_{money} \left[\frac{dV_{Oa}}{V_{Oa}} - \frac{dV_{Oc}}{V_{Oc}} \right] \qquad (4.32)$$

and

$$ds = k_{money} \left[\frac{dG_{Oa}}{G_{Oa}} - \frac{dG_{Oc}}{G_{Oc}} \right] \quad [subject\ to\ \textbf{dP/P}\ active = \textbf{dP/P}\ inactive]$$

$$(4.33)$$

Where k_{money} = 1 (£1, $1 etc)

Incremental money entropy change was related to the rate of change in *active* output volume flow, moderated by the rate of change in the *inactive* part, with a similar picture for value flow change, *provided that the rates of change in price were the same for both active and inactive parts.* Economic entropy change can be positive or negative, depending on the changing relationship between *active* and *inactive* parts. It has elements of rate of change per unit of time.

While the author does not propose that economic entropy conceived in this manner can be formally equated with that arising from the consumption of the real productive content/exergy of resources, labour and capital stock, because efficiency losses and waste are excluded from economic accounts, it does conform to the same laws of thermodynamics, and does affect the way in which economies develop. However, before proceeding further with our development, it is necessary to digress a little to the concept of *order* versus *disorder*, as understood by physicists, and to explain what this means in economic terms.

Most people would accept that a walled garden, left to its own devices untouched by the outside world *[a relatively closed system, bar weather and visiting vegetable and animate matter]*, would, over time, gradually become less ordered – weeds would grow, bits might crumble off the garden gnome and the spade left out in the open might go rusty. Leave a bucket of hot water out in the open and it soon gets cold *[unless it was in the middle of the Sahara desert – an open system subject to the suns fierce rays]*. It's an irreversible process. The universe and all its contents over time become more disordered with temperature decreasing, conforming to the Laws of Thermodynamics, with entropy forever increasing.

It does not take much thought, however, to realise that against this disorderly process there runs a process in the opposite direction *creating* order. Animals and plants are ordered entities, compared to say a pile of stones or a heap of rubbish. So the argument might be put that doesn't the order generated compensate for the disorder? After all, the world is full of living things, humans and the trappings of mankind, which are all highly ordered entities.

The simple answer to this is no, for in order to create an ordered state with a lower entropy level, according to the Laws of Thermodynamics, more energy has to be consumed to increase the order level than is actually transferred to and contained in the increase in order. In other words, some exergy value is thrown away, that cannot be retrieved. Mahulikar and Herwig (2004) state that, in an isolated system, order created within disorder is a state of *higher* net disorder than the preceding state of total disorder. Order can only create more order through *disorder*, and the role of creation of order therefore is to increase the system entropy at a faster rate than had order not existed. The chart at figure 4.8 summarises the effect.

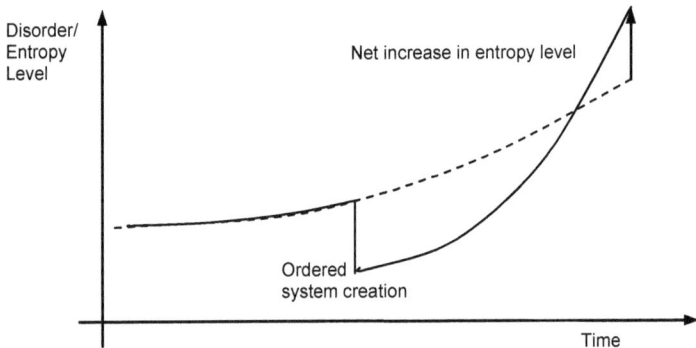

Figure 4.8 Disorder over Time

To give an economic example, suppose a man lives in the open where it is cold and inhospitable. He decides to build a house to improve his well-being. So he corrals together, through a complex production process, all the things he needs and builds the house, expending energy and consuming inputs to create something lasting. He has just ordered things for his benefit *[and created some value which he could sell if he wished]*. Houses, being what they are however, gradually fall into disrepair, unless given some love and attention, and in the ultimate, if left for a long time, decay and fall down, losing their value. There are few remains and relics of former ages of man. Thus not only does the house gradually lose its value over time, but some energy has been expended and lost, over and above what went into the house. In other words, more entropy has been generated than would have occurred if the house had not been built in the first place. Thus, in the ultimate, the ordered system is just a more efficient means of increasing the rate of creation of disorder over time.

We are therefore back to Schrödinger's *'order from disorder'* premise (1944), mentioned in chapter 1, which was an attempt to link biology with the theorems of thermodynamics, whereby a living organism maintains itself stationary at a fairly high level of orderliness (low level of entropy) by continually sucking orderliness from its environment. Annila and Salthe (2009) echo this thought, regarding economic activity as being an evolutionary process governed by the second law of thermodynamics.

The link that is missing therefore is the principle or algorithm that determines the choice from among many options that a decision maker takes to increase the entropy level.

Swenson (2000) has noted that while the Second Law says that thermodynamic systems act to minimise potentials and maximise entropy production, it does not state which path, out of a choice of paths, such systems will take to do this. He has proposed a Law of Maximum Entropy Production which states: *'A system will select the path or assemblage of paths out of available paths that minimises the potential or maximises entropy production at the fastest rate given the constraints.'* An example is provided by Swenson, whereby the door to a log cabin *[which contains some residual heat]*, located somewhere in a cold wood, is suddenly opened. Instead of heat and hot air seeping out only through the walls and cracks of the log cabin into the cold air of the wood as it did before, it automatically selects the open door to exit the building, increasing the rate of cooling in the log cabin, until equilibrium is reached when the temperature inside matches that outside. Heat does not stop flowing out through the cracks; just that most of it now goes out the door, and at a faster rate.

At a higher level, plant life and grazing animals follow a similar process, continually seeking out the shortest route in terms of time to a source of water and nutrient energy, but if this subsequently dries up or is not renewed by nature and the natural cycle, they spread their net further to seek other sources. If there are no further reachable resources they will eventually wither and die. Effectively life endeavours to follow a path of maximum entropy gradient, choosing the first available path in terms of time that does this, but if this subsequently proves not to be fruitful or provide further opportunities, it seeks to choose another path.

In the ultimate, human life also follows such a process, though humans, with a higher level of gene and intelligence, are able to make use of the resources that the earth has to offer in a manner, quantity and speed that no other (earthly) species can match, and can also consider longer timescales. More recently in the ecological timescale, humankind has developed from requiring only basic food, shelter, and breeding/germination as other animals and living organisms, to engineering a significant diversion of nature's resources to fulfil particular and developing desires not strictly necessary for the preservation of the species. A further *ordered* consequence has been the mushrooming of the human population and its consumptive manner of living compared to some other species.

The inference is that economic systems, being the brainchild of humankind, also operate in a manner to maximise entropy production, enabling humankind to acquire a more ordered state for itself, at the expense of increased disorder in the ecological system.

Drawing together the threads between economics and thermodynamics, we can see that the underlying principle or algorithm is one of maximisation of potential economic entropy gain. Decisions that are favoured include:

- Pursuing positions where the activity rates of *inactive* inputs can be increased easily and with minimum cost.
- Investment in economic structure to change the system to supplant/replace a constraining component with higher net yielding output; for example, power from energy sources replacing manpower.
- An emphasis on shorter term projects compared to longer term ones, as the quality of information concerning the future decreases with time.

The impact of this process will be examined in later chapters. Chapter 5 deals with money and the interaction of interest rates with the money supply and the economy. Chapter 6 deals with employment and unemployment, and chapter 7 sets out the schematics of discounted cash flow analysis and bond investments and yield. And last, chapter 8 applies the same logic to energy demand and supply, with a further extension to climate change.

The conclusion of this section is that it is reasonable to propose that a thermo-economic system will seek to maximise economic entropy gain over time (thereby potentially increasing main-body entropy production), but subject to the relationship between the *active* and *inactive* parts of the economic system. It is emphasised that the paths of potential economic entropy can be moderated by one or more of a whole series of constraints, sometimes impacting at different times, producing irregular patterns of output flow.

We are thus dealing with systems that are forever dynamically changing and not in a state of equilibrium, a form of non-equilibrium economics, a topic that has been the subject of other researchers. Chen (2002) has investigated economic and biological systems as open dissipative systems extracting low entropy from the environment to compensate for continuous dissipation; and Martinas (2006), investigating non-equilibrium economics, concluded that perfect competition of selfish agents does not guarantee the stability of economic equilibrium, and that the role of economic equilibrium is not justified by thermodynamics.

4.4   The Cycle

Physicists and engineers commonly represent the workings of thermodynamic systems by reference to a cycle of operations. Such a cycle is composed of several stages. For example, in a fossil-fired engine, first fresh air is introduced and then compressed. Then fuel is injected and ignited in the air to give off a gas (CO_2) and heat. The heat produced expands the gases, which generates work in excess of that required for compression. Finally, residual gases and heat value are expelled, ready for the next cycle. Such cycle operations generally occur over very short periods of time – engines in cars and steam & gas turbines for example typically turning over at several thousand revs per minute, many times a second.

It might be supposed that similar processes could be imagined for an economic system, accommodating the production and stock processes set out so far in this book, with reactants of economic input being brought together and transformed into economic output that is subsequently returned as input – a cyclical process. Generally, however, such cycles takes place over a range of mostly longer periods than thermodynamic ones; up to five or more years for some investment projects, down to seconds and minutes for small consumer products.

High output in an economic sense is predominantly achieved by large numbers of people and machines acting together, at different times, alongside or by interaction with each other. Most advanced economies have quite a few million working people, and a typical working year for each person might involve say 240 days, 150,000 minutes per year. Thus what occurs in a macro-economic context is an 'average' of many machines and people working as a 'tumult'.

In economic theory, the term cycle is traditionally referred to in terms of the 'business cycle', generally occurring over periods of years, with the root causes of variations over the cycle period being attributed either to *external* factors (changes in population growth and resources, wars, technological factors and even sun spots), or to *internal* factors (psychological, political, investment changes and the acceleration principle, Keynesian economics). In this book, the primary description of such a cycle is couched in terms of money and changes in economic entropy and elasticity, which will be described in the next chapter.

While the author has considered a thermodynamic approach to cycles in the past, there are however several problems, concerning the differences between economic and thermodynamic systems, that make the idea of a thermo-economic cycle, as set out by engineers, difficult to conceive.

A number of thermodynamic cycles have been devised to describe the dynamics and efficiencies of thermodynamic systems, but that devised by the French physicist Sadi Carnot (1837-94) is recognised to have the maximum theoretical efficiency that can be obtained. The cycle is made up of two isothermal processes and two isentropic processes, as in the **P-V** and **S-T** cycle diagrams at figure 4.9.

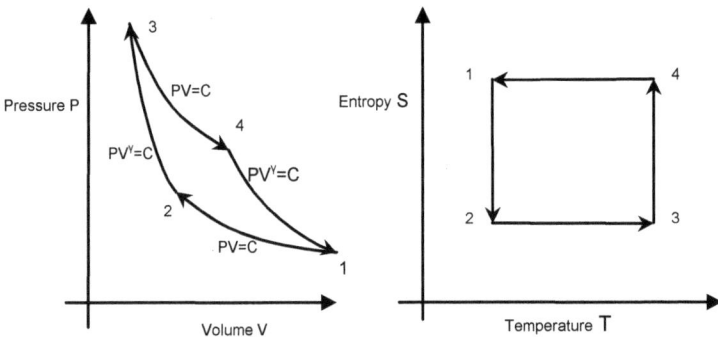

Figure 4.9 Carnot Cycle

In the cycle, gas inputs are compressed at process 2-3 consuming work value. In process 3-4 they combust with fuel and acquire heat value, and at process 4-1 they expand giving off work value. Last, at process 1-2 heat value is lost to the ecosystem as residual gases are expelled. For a number of reasons, but chiefly because the area inside the **P-V** diagram is very small, implying little net work done, the cycle is not a very practical one. To overcome the problems posed by the Carnot cycle, more practical cycles in use by engineers include the *Joule* cycle [James Joule 1818-89], the *Otto* cycle [Nikolaus Otto 1832-91], and the *Rankine* cycle [William Rankine 1820-72]. In the Joule cycle, for instance, constant pressure processes replace the isothermal processes of the Carnot cycle.

The first difficulty met with economic systems is that they are composed of several intertwining cycles – human, money, resources, and natural cycles, which all have very different cycle periods and feedback systems. They are also composed of very different 'fluids', people, money, and natural substances - compared to a gas thermodynamic cycle. Thus there is a significant problem of homogeneity and cycle times, though this is accommodated by the use of money as a means of exchange.

The second, rather more intractable, problem is that economic systems have elements of both 'flow' and 'non-flow' components, relating to time. Gas systems, on the other hand, can be separated out, either flow or non-flow; and while the flow system contains time, this can be conveniently eliminated in respect of P-V and S-T diagrams. Figure 4.10 shows how engineers achieve this.

The time element introduced into the flow system is conveniently side-lined by dividing volume flow V_t by gas molecule flow N_t to produce a dimensionless specific volume **v**. Likewise, entropy flow level S_t is divided by gas molecule flow N_t to produce an entropy level **s** per unit of gas. Thus in both sets of diagrams time has been eliminated.

Such an arrangement cannot be set out for an economic flow system, because, although time **t** is allocated to the volume flow **V**, the stock volume **N** does not have a time dimension, and consequently this is transferred to the index of trading value **T**, as shown at figure 4.11.

Non-flow Thermodynamic System

Flow Thermodynamic System

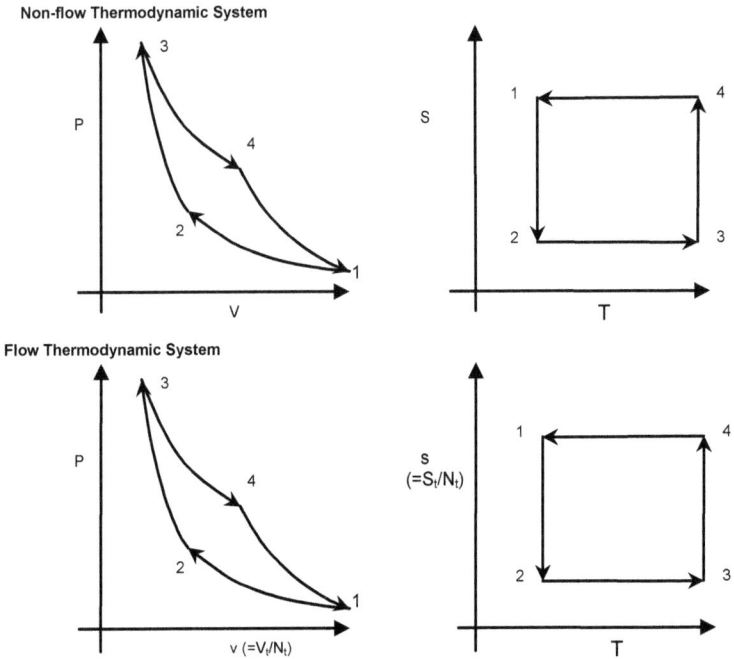

Figure 4.10 Engineering Flow and Non-Flow P-V and S-T Diagrams

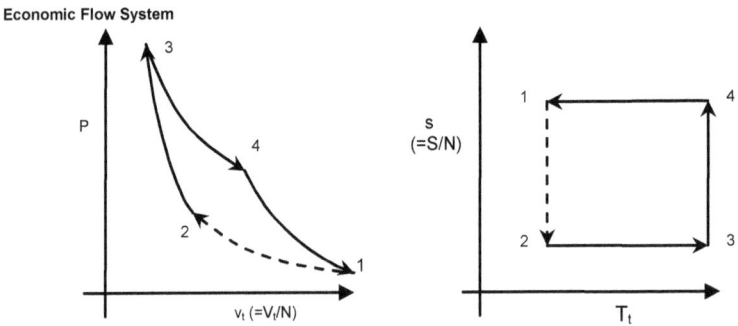

Economic Flow System

Figure 4.11 Conceptual Economic Flow P-V and S-T Diagrams

A third problem is that while thermodynamic systems allow for losses in efficiency to be calculated, this cannot be done for economic systems, since no losses are set out; for example, heat losses up a power station chimney. Such waste is not given a value, and added value in is equated to added value out. Economics operates a strict 100% efficiency ratio, and does not allow for the substantial losses that nature picks up. Thus process 1-2 at figure 4.11 does not have a value in an economic system and the cycle is in effect 'broken'. As a consequence a concept such as overall cycle efficiency does not have much meaning in economic analysis.

Finally, it will appreciated that once economic value is aggregated at above the micro-level, the concept of a difference between inputs at process 2-3 and outputs at process 4-1 becomes progressively reduced, as in table 4.1.

Process	Micro	Macro
• 2-3	Bought-in costs	Imports
• 3-4	Added Value	Domestic Added Value
• 4-1	Sales Value	Exports
• 1-2	Waste (no value)	Waste (no value)

Table 4.1 Micro and Macro Descriptions

At the macro level, imports can become equal to exports, and domestic added value effectively comes out of the ecosystem, though it is attributed to humans as a wage/profit.

For all of the above reasons, the author considers that more research is required in this area to make the approach meaningful.

CHAPTER 5   MONEY

The thermodynamic ideal economic equation in the opening chapter was set out on the basis of a series of interrelated stocks, of which the money stock was central, serving all the other stocks. Value and output flowed around the economic system, interacting at every stage with the banking system.

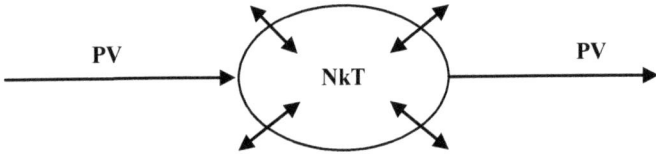

Figure 5.1 A Money Stock

For a money system, the ideal economic equation was expressed as in equation (2.4):

$$P_M V_Z = N_M k_M T_M \qquad (5.1)$$

Where P_M is the price level, V_Z is output volume, N_M is the number of currency instruments in circulation, k_M is a *nominal* monetary standard (£1, €1, \$1, etc) and T_M is an Index of Trading Value, here the velocity of circulation of money.

This equation is a re-statement of the general quantity theory of money. The quantity theory is commonly written as PY=MV, where P is price level in an economic system, Y is output in volume terms, M is the quantity of money in circulation (variously defined as M0 – M4, but with M2 & M4 generally being recognised as standards), and V is the velocity of circulation. While the left-hand sides of the equations are comparable, the right-hand side requires additional clarification. To obtain an exact comparison with the quantity theory, equation (5.1) can be written as:

$$PV = [Nk]T \qquad (5.2)$$

Where **Nk** is equivalent to money M in circulation and the index of trading value **T** in a thermodynamic monetary system is equivalent to the velocity of circulation V in a traditional monetary system. To avoid any confusion concerning the use of algebraic symbols, the quantity theory equation PY=MV is dispensed with in this book. The equation has a very similar format to the ideal gas equation. Pikler (1954) has highlighted the connections between the velocity of circulation and temperature.

Thus, in equation (5.1), there are four variables to consider, and in differential form equation (5.1), from now on without the subscripts, can be written as:

$$\frac{dP}{P} + \frac{dV}{V} = \frac{dN}{N} + \frac{dT}{T} \tag{5.3}$$

The fifth factor, the nominal currency value k, is fixed by definition, though it can change its effective value through inflation and international comparison.

At this stage we are *not* pre-forming a view on how the factors inter-relate, and whether they follow a Monetarist, Keynesian or other school of thought. All that can be confirmed is a tautological identity, with a change in one or more of the factors automatically equating to a balancing change in the others, in a manner so far undefined. All of the factors can vary, and none can be considered to be constant. A particular point to note with the model so far outlined, however, is that the velocity of circulation T, calculated as output in value terms divided by money supply, carries a connotation of value too, since the nominal value k is *deemed* to be unchanging, and the number N of monetary instruments is just that − a number. If the number of monetary instruments N in circulation remains constant, when output price P and output volume V are each changing in some fashion, then the changes in P and V are reflected in a change in the velocity of circulation T. T can therefore carry value as well as volume. This is not to say of course that the number of money instruments N necessarily remains the same.

Figure 5.2 illustrates the trends of economic development of the UK and USA economies. UK data is taken from quarterly statistics of Economic Trends Annual Supplement (www.statistics.gov.uk). USA data is taken from quarterly statistics of www.federalreserve.gov and www.bea.gov. Volume data is taken from GDP chain volume measures *[UK 2008 prices, USA 2005 prices]*. The Price deflator is calculated by dividing GDP at market prices by the volume measure. The raw data was smoothed by calculating 4-quarter moving averages. The date range was decided by what data was easily available. UK figures are based on an M4 definition of money, whereas US figures are now based on M2 (publication of M3 figures ceased in 2006).

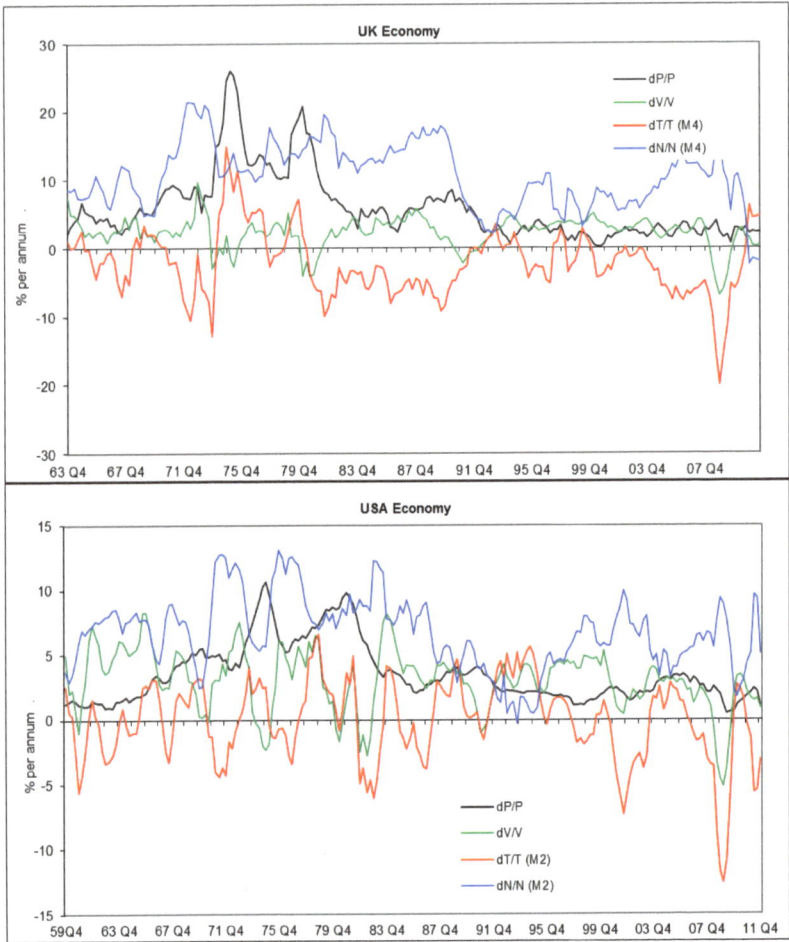

Figure 5.2 Annualised 4-quarter moving averages of percent growth rates of key economic variables. GDP Deflator (P), Output Volume (V), Money Stock (N) [UK M4, USA M2], and Velocity of Circulation (T)

A particular problem with a money system, compared to an energy system, is that whereas in the latter the value **k** is defined as a constant, being energy per molecule per degree of a scale of temperature, which can be derived by experiment, in a money system, inflation can have the effect of reducing the effective value of money. The nominal fixed amount **k** remains the same, but the effective real value of money can decline. However, when viewed from the point of view of the money instrument **k,** it sees itself as unchanging (£1, €1 etc), but all around it is changing. History abounds with attempts to derive an unchanging standard, such as gold, man-hours and energy content.

A less dramatic but still important issue is that of the definition of output volume **V**, given that over long periods of time the mix of goods and services changes, entailing sophisticated deflator estimates to try to strip inflation out and take into account switches from high volume low added value to low volume high added value. Last, calculation of the velocity of circulation is by definition a residual exercise, being equated to total output at market prices divided by total money instruments, rather than being set against an independent scale. Technical changes in the usage of money can also occur, altering the velocity of circulation. The combination of these problems makes for a more complex analysis.

5.1 Development and Elasticity

Before developing a thermodynamic representation of a money system, a point should be made concerning the number of variables. In a *non-flow* thermodynamic system, both the constant **k** and the number of molecules **N** are fixed, and there are therefore only three variables left to consider, pressure **P**, volume **V** and temperature **T**. In a thermodynamic *flow* system, on the other hand, there are four variables left (other than the constant **k**): molecular flow **N**, volume flow **V**, pressure **P** and temperature **T**, but these are effectively reduced to three, by replacing volume flow **V** and molecular flow **N** with specific volume $v = V/N$. Thus again only three variables remain, enabling thermodynamic analysis to proceed.

To develop a thermodynamic analysis of a money system, a similar requirement is necessary; that is, to reduce the four factors in equation (5.3) plus **k** (making five) to three, preferably without losing output volume flow **V**.

In chapter 3 thermodynamic principles, when a *non-monetary* good was considered, the productive content **k** of each unit of good was considered to be constant, and volume flow **V** was divided by stock number **N**, to derive a specific volume flow rate **v**. The latter, combined with price **P** and the index of trading value **T**, then constituted only three variables, which enabled the thermodynamic analogy to be constructed. However, in a money system, all five factors can vary. Although the money value **k** is constant (i.e. £1, $1), this is only thus in *nominal* terms. A period of extensive inflation can reduce its effective value. Thus dividing a real output flow **V** by the money stock number **N** still leaves four variables. Moreover, dividing a volume output by a number of units of a depreciating currency might negate the principle of defining output at constant prices. As the relative level of volume output per unit of time also defines to a large extent the size of an economic system, preserving the integrity of volume flow **V** is important. By way of illustration, in the UK between 1963 and 2011, money stock (M4) grew by a factor of 153, output volume by a factor of 3.1 and prices by a factor of 16.9; in the USA between 1959 and 2011 money stock (M2) grew by a factor of 33.4, output volume by a factor of 5.0 and prices by a factor of 6.2. Compensating adjustments in the velocity of circulation accounted for the remainder. It is inconceivable that technical change adjustments equated to inflation.

A possible part solution to this problem is to transpose the price deflator **P** to the right hand side of the equation, to derive a *Specific Money Stock* N_P= **N/P**, leaving output volume **V** alone on the left hand side. This arrangement, though acceptable, does not accord with standard presentations of thermodynamic analyses, or of general economic practice of having both price and volume flow on the output side of the equation. A subsidiary problem to consider, however, is the extent to which changes in price **P** reflect loss of currency value **k**, or gains in real value. This all rather depends upon from where an observer is standing in order to view an economy and its surrounding economies. All is relative, though some factors might be more relative than others. Nevertheless, the net effect of using this method will likely be to understate money entropy change and this point should be born in mind all through this chapter.

The probability is that any change in money instrument stock **N** may find its way into changes in all three of the other variables, price, output volume or velocity of circulation, depending upon the relative elasticity between the three and with money, and the degree to which a money system is out of kilter with the stable state, such as the existence of excess money or high inflation.

In this book, therefore, the preferred arrangement to develop the thermodynamic characteristics, although not perfect, is to transpose money units **N** to the left-hand side of the equation, and divide output price level **P** by the number **N** of monetary units in circulation to give a Specific Price P_N(=**P/N**). Presentations of the inverse N_P (= **N/P**) are also given at each stage however for completeness, as the results are the same.

Thus equation (5.1) could be written as:

$$P_N V = kT \tag{5.4}$$

Or in the alternative:

$$V = N_p kT \tag{5.5}$$

By differentiating equation (5.4) and dividing by $P_N V = kT$ we have:

$$\frac{dP_N}{P_N} + \frac{dV}{V} = \frac{dT}{T} \tag{5.6}$$

Where (dP_N/P_N) also equals $-(dN_P/N_P)$. Figure 5.3 illustrates the relationships between the factors.

A change in velocity of circulation T can arise either from a change in specific price P_N or a change in output volume V, or both. And no change in the velocity of circulation T will occur if a change in specific price P_N is balanced by an equivalent change in output volume V.

As illustrated at figure 3.6 in chapter 3, a number of relationships between price and volume are current in a thermodynamic presentation. Figure 5.4 illustrates the same relationships, but in terms of specific price P_N.

For an economy where no change in output volume occurs (V = **Constant**), increases/decreases in specific price P_N are matched by an equivalent change in the velocity of circulation T. Such changes in specific price P_N involve a change in the relationship between actual price level P and the money stock N.

At the other extreme, where no change in the specific price P_N occurs, changes in real output volume V are matched by appropriate changes in the velocity of circulation T. Should a movement in money supply N occur, this will be balanced by a change in price P.

If the velocity of circulation T remains constant, then any change in output volume V will result in an offsetting change in the specific price P_N, accompanied by a change in the relationship of price P to money supply N.

124

Last, a further relationship between the variables can exist, which in this book we have called the Polytropic case ($P_N V^n$ = **Constant**), involving an Elastic Index **n**, where changes in all of the factors can take place, but in a complex manner. This is shown at figure 5.4.

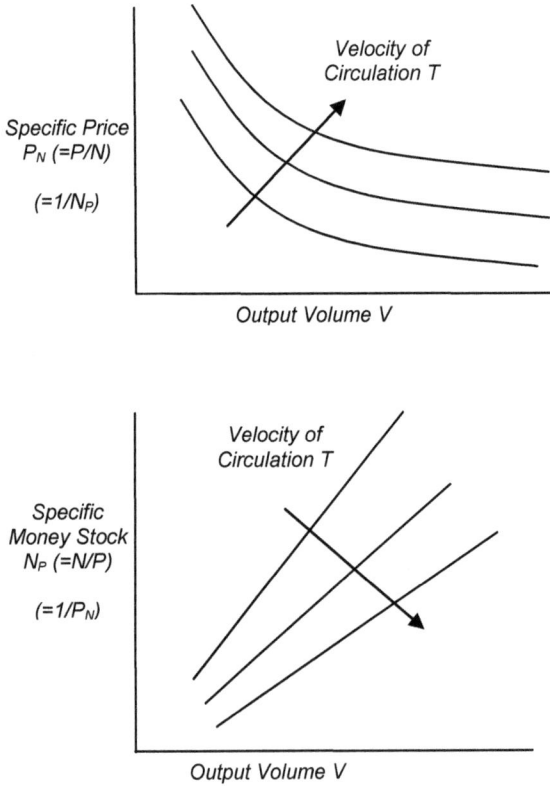

Figure 5.3 Velocity of circulation, specific price, specific money stock and output volume

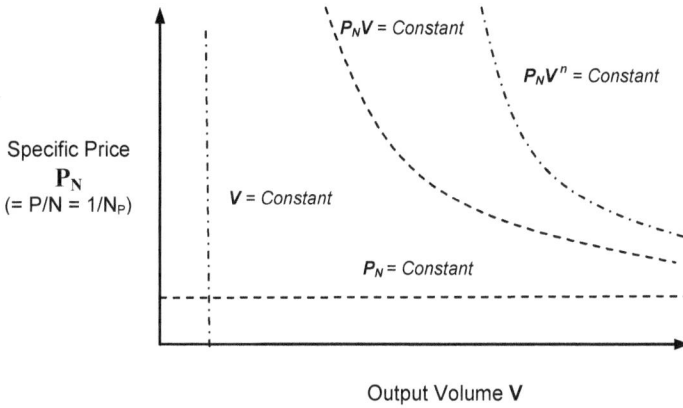

Figure 5.4 Specific Price – Output Volume relationships

The polytropic equation $(P_N V^n = \text{Constant})$ can be adapted to meet all of the possible processes. A constant specific price process $(dP_N/P_N=0)$ for example is a polytropic process with the elastic index **n** set at zero, and a constant volume process is one with the elastic index set at $\pm \infty$. The iso-trading model $(P_N V=\text{Constant})$ has an index of 1, with no change in velocity of circulation **T**.

Because the monetary stock system is effectively *one* stock system, though serving all the other sectors, we could use the polytropic case to describe its dynamics, since this covers all the other price-volume relationships.

By combining equation (5.6) with the polytropic equation $(P_N V^n = C)$, the following equations describe the polytropic case:

$$\frac{P_{N2}}{P_{N1}} = \left(\frac{V_2}{V_1}\right)^{-n} \qquad \frac{dP_N}{P_N} = -(n)\frac{dV}{V} \qquad (5.7)$$

$$\frac{T_2}{T_1} = \left(\frac{P_{N2}}{P_{N1}}\right)^{\frac{n-1}{n}} \qquad \frac{dT}{T} = \left(\frac{n-1}{n}\right)\frac{dP_N}{P_N} \qquad (5.8)$$

$$\frac{T_2}{T_1} = \left(\frac{V_2}{V_1}\right)^{1-n} \qquad \frac{dT}{T} = (1-n)\frac{dV}{V} \qquad (5.9)$$

Thus far, we have not specified how the relationships impact on the way in which an economy moves and what drives the relationships, only that changes in one or more of the factors will be reflected by changes to the others to enable the equation of state set out at equations (5.1), (5.3) and (5.6) to balance out.

Figures 5.5a and 5.5b illustrate the trend of the relationships for the UK and USA economies, making the adjustments to $P_N = P/N$ (and $N_P = N/P$), as per equations (5.7), (5.8) and (5.9). A trend line drawn through the curve at figure 5.5a for the UK economy is of the form $P_N V^{1.95} = 1.23$, with a regression of $R^2 = 0.97$. This result, however, masks the significant disturbances brought about by the large inflationary boost in the period 1972 to 1982. There was also a significant reduction in the velocity of circulation. The onset of recession from 2008 onwards also brought about substantial disturbances to the trend, shown up particularly by the path of N_P in the lower chart.

A trend line at figure 5.5b drawn through the curve on the basis of M2 figures for money supply for the USA economy is of the form $P_N V^{0.894} = 0.63$ with a regression of $R^2 = 0.98$. This result, however, masks a 5-year change in the relationship 1989 to 1994 when growth in money supply N declined and velocity of circulation rose, before settling back to its original slow decline. As with the UK economy there have been significant disturbances, brought about by the recent recession from 2008 onwards.

From 2006 onwards, figures of M3 money supply in the USA have not been produced, and the authorities now rely only on the narrower M2 definition for economic management purposes. Figure 5.5b shows that the trends of the two sets of data follow a similar path, though gradually diverging with time. By 1st quarter 2006, USA M3 money supply figures had grown to a level of 53% above M2, from being nearly the same in 1959.

The above regressions for the two economies of course average out short term variations.

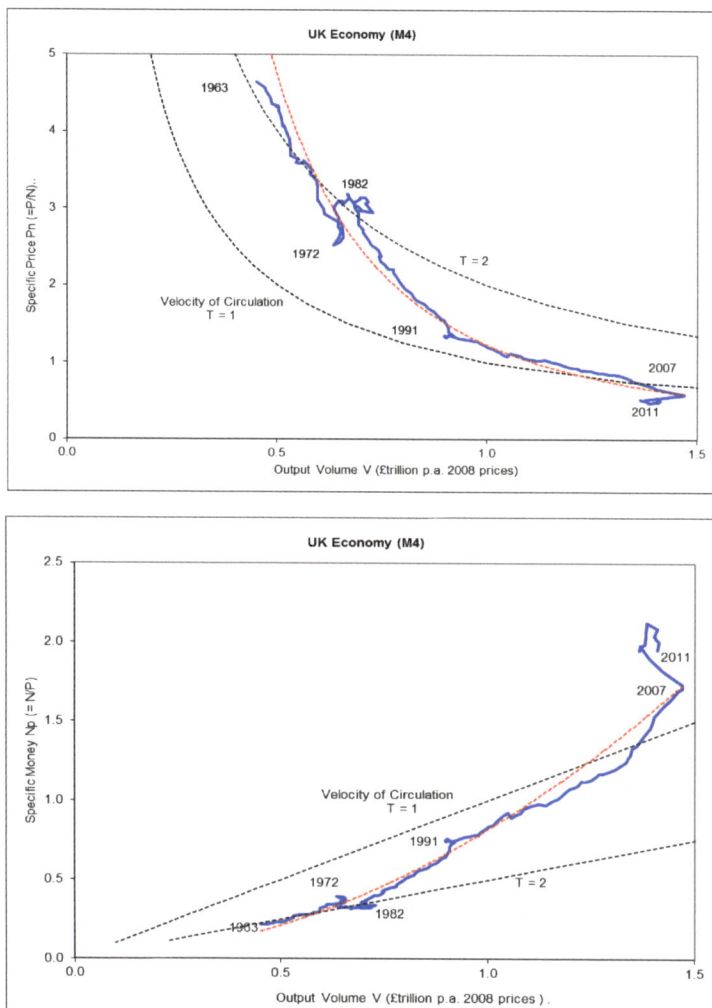

Figure 5.5a *Quarterly data Specific Price, Specific Money, Output Volume and Velocity of Circulation – UK*

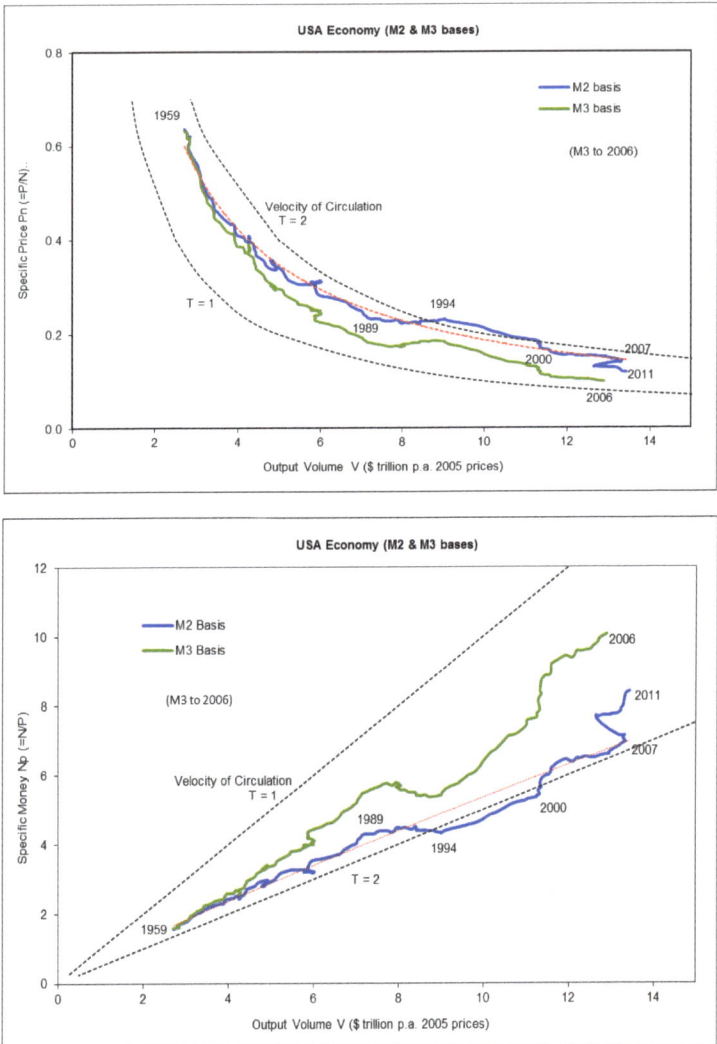

Figure 5.5b Quarterly data Specific Price, Specific Money, Output Volume and Velocity of Circulation - USA

Because price **P** has been divided by money stock **N** to compute a specific price P_N, and vice-versa for specific money N_P, the relative elasticities between the variables in the polytropic case will be different compared to the position with these variables being separate. Nevertheless, the above results indicate that elastic relationships of specific price P_N and specific money N_P with respect to output volume V have existed in both the UK and USA economies over much of the periods. Examination of data of M0 money, rather than M4 or M2, yields different relationships between the variables. However, according to Harrison [Harrison et al Bank of England (2007)], M0 is now very small and forms only 3% of M4.

Figure 5.6 sets out the short term development of the annualised rates of change of the variables **dV/V, dT/T and dPn/Pn** for the two economies. The initial data was smoothed by calculating annualised 4-quarter moving averages.

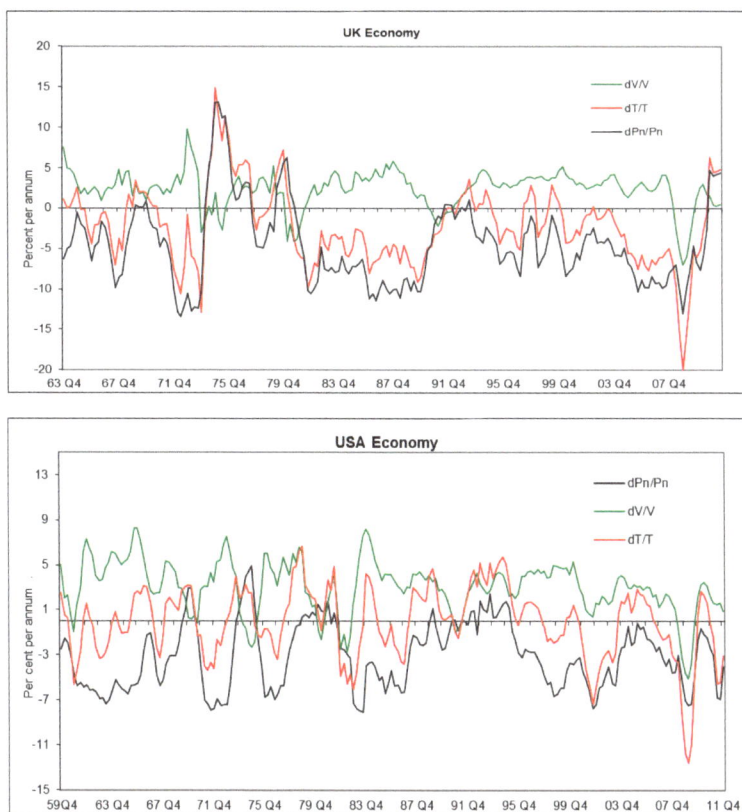

Figure 5.6 Annualised 4-quarter moving average percent change in Output Volume V, Specific Price P_N and Velocity of Circulation T.

By utilising the differential form of equation (5.7) it is possible to calculate the short-term quarter-on-quarter variation in the elastic index **n** for the two economies.

$$\frac{dP_N}{P_N} = -(n)\frac{dV}{V} \qquad \text{and} \qquad \frac{dN_P}{N_P} = (n)\frac{dV}{V}$$

Figure 5.7 sets out the variations in the elastic index **n** for the two economies on this basis, using the same smoothed data used to present the trends at figure 5.6.

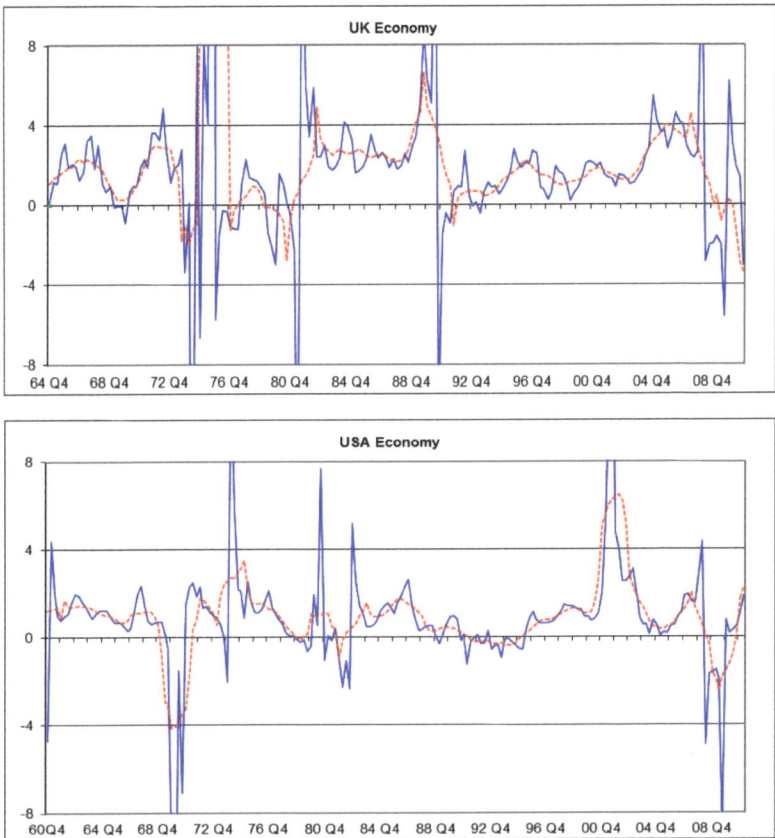

*Figure 5.7 Elastic Index **n** of the UK and USA Economies (with centred 8-quarter moving average)*

Some of the data exhibited wild swings outside the scale of the charts, owing to some points where changes in specific price P_N were divided by very small changes (positive or negative) in output volume V *(see figure 5.6)*. An explanation of this effect is given by reference to figure 5.4. For processes operating at approaching either side of constant output volume **(dV/V=0)**, the elastic index can approach a very high number (plus or minus). Figures 5.6 and 5.7 illustrate points in the two economic cycles where the effect occurred. Readers may note also that these points occurred at particular 'recessive' years, some of which coincided with periods of turbulence in world oil markets. It can be seen that movements in the elastic index are the 'norm' as an economy develops.

The three charts at figure 5.8 illustrate the effect of a change in the elastic index **n** between specific price P_N, output volume V and velocity of circulation **T**.

To finish this section, we refer back to the polytropic equation (5.7). Splitting this into its component parts, we have:

$$P_N V^n = \left(P/N\right) V^n = Constant \qquad (5.10)$$

And taking logs and differentiating we have:

$$\frac{dP}{P} = \frac{dN}{N} - n\frac{dV}{V} \qquad (5.11)$$

Figure 5.9 illustrates actual annualised quarterly changes in the price level change **dP/P**, set against projections of price change using equation (5.11) above for values of elastic index **n** of 1.5 and 1.0 respectively for the UK and USA economies, taken from the results following the charts at figure 5.5 of this book. The projections follow the lines of change in price, but wildly under and over-shoot, because account of the short-term change in the elastic index **n** by quarter has not been taken.

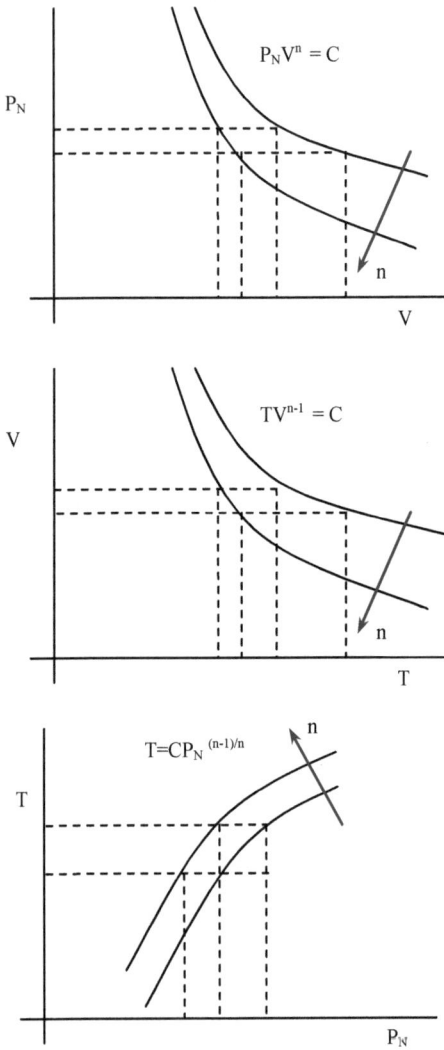

*Figure 5.8 Elasticity between Specific Price P_N, Output Volume V
and Velocity of Circulation T*

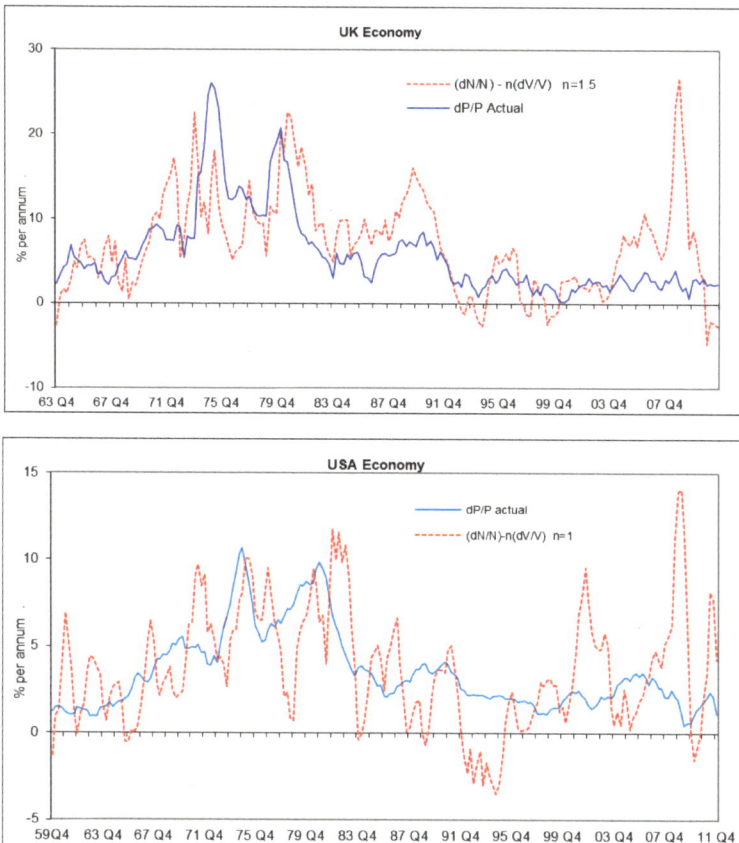

Figure 5.9 Projections of percent price change for the UK and USA economies using a thermodynamic equation (5.11), set against actual annualised quarterly changes

5.2 Money Entropy

Monetary systems are subject to variations in entropic value, just as the resources passing through them are. Such variations in money entropy as occur, however, will be confined only to preservation of the exchange value of money balances. They will *not* include entropy changes arising from efficiency losses incurred through consumption of the productive content of resources to produce output; the latter occur through the normal mode of flow of product value through an economy. Money has no value except as a means of exchange between items of real value. On a normal basis therefore, money entropy operates *at the margin*, with monetary authorities acting mostly to preserve currency values. From the development at chapter 4, readers will appreciate that money entropy (in particular equations 4.23-4.25) has been developed to be a measure of the motivating force in an economic system.

From all the foregoing, entropy value in a monetary system might therefore be described as *potential* entropy, though a real effect might be the complete loss of value from printing money. A consumer in a poorly managed rogue economy, armed with a pile of money that only a day or so previously merited an exchange value of a certain amount, might find subsequently that he could not exchange the money for anything. Clearly the nominal value of the money had suddenly experienced a complete exit of entropic value; very real for the man involved, entailing a loss of purchasing power of items of real value which do have a real useful productive content with a realisable entropy value. Despite the loss of economic entropic value, the products and services that our man had already claimed ownership of do not contain anymore or less productive content than they did before. They would loose their inherent productive content over time in the normal entropic way, though perhaps our man might try to conserve their use! If the man had traded on a barter basis instead, he might have retained his value.

The analysis so far indicates that the UK and USA economies have followed polytropic paths, if price level **P** is replaced by specific price $\mathbf{P_N}$, equal to price level **P** divided by money instruments **N** (M4 and M2 money basis); and vice-versa if money instruments **N** are replaced by specific money $\mathbf{N_P}$, equal to money instruments **N** divided by the price level **P**. It will be appreciated that movements in the elastic index **n** are continually occurring, with wild swings when volume change is small or turning negative.

From the Second Law of Thermodynamics, the net change in entropy per unit of stock through any cycle is stated as:

$$\Delta s_{cycle} = \oint \frac{dQ}{T} \geq 0 \qquad (5.12)$$

Entropy through the cycle tends to rise. Moreover, a consequence of dividing price by money units to compute a specific price $\mathbf{P_N}$, in order to get round the problem of a depreciating currency, is that entropy increase may be understated. Nevertheless, excluding resource exergy losses, monetary economic cycles by themselves can be quite efficient, for if consumers agree to buy products and services from suppliers with an agreed productive content between the two parties, but with low inflation, one might expect only a small rise in money entropy during the exchange. Once consumers have bought their product/service, however, then a large increase in entropy occurs, as the products go through their useful life with consumers, or are consigned to waste. This part of the cycle is not efficient, and is to be regarded as a Second Law loss.

A special case of the polytropic form $(P_N V^n = \text{Constant})$ is the Isentropic case where entropy change through the process is zero. It has the form $P_N V^\gamma = \text{Constant}$, where the index γ is a constant. Thus the structure for an isentropic case remains the same as that for the polytropic case (equations (5.7) and (5.8)), but with the elastic index n replaced by another index γ.

While in our monetary process the index γ is unknown, clearly, from figure 5.7 of this chapter, the elastic indices n for the UK and USA economies have followed significantly variable paths, and isentropic conditions, have not existed over the periods examined.

It was shown in chapter 3 that an expression stating the change in entropy in a polytropic process could be set out as in equation (5.13) below. While this held for a single unit of stock $(s=S/N)$, we have effectively also arranged the same here by dividing price by money stock $P_N=P/N$). Thus:

$$ds = k\left(\omega + \frac{1}{1-n} \right)\frac{dT}{T}_{rev} \qquad (5.13)$$

Where the expression in the brackets was called the Entropic Index λ. The entropic index λ was related both to the elastic index n and the value capacity coefficient ω, which represented the amount of value required to raise the index of trading value (the velocity of circulation) T by a given increment, here for a nominal currency value k of 1. The relationship of the entropic index λ to the elastic index n and the value capacity coefficient ω with respect to changes in the velocity of circulation T is shown at figures (3.13), (3.14) and (3.15) at chapter 3. Remembering that for a polytropic case (equation (5.9) the following equation applies:

$$\frac{dT}{T} = (1-n)\frac{dV}{V} \qquad (5.14)$$

Then an alternative expression for the entropy change is:

$$ds = k(\omega - \omega n + 1)\frac{dV}{V}_{rev} \tag{5.15}$$

It was shown at equation (3.24) that the value capacity coefficient ω is a function of the *lifetime* length and perhaps other unquantifiable aesthetic factors. In the case of money, however, aesthetic factors are unlikely to be of significant importance and the lifetime length is likely to be the key factor. For money in the form of cash or electronic transfers, ω will be short – much less than a year. For money in the wider M2 or M4 definition, ω will be longer, and will be an average of a number of different constituents, up to 5 years. Thus ω is dependent upon the nature of the money instrument. While ostensibly for a given mix of money instruments ω might be fixed, it should be noted that changes in money mix, for instance a move towards electronic money, can change this value. From the charts at figure 5.5, the velocity of circulation of the UK and USA economies has varied about 1-2 times a year, suggesting a lifetime value for M2 and M4 money in the region of $0.5 - 1$ year, with a value capacity coefficient ω also at about this level. A wider definition of money would entail a longer lifetime. In the UK, the lifetime appears to have lengthened from $1963 - 2011$ as the velocity of circulation has reduced. In the USA, the lifetime of M2 money appears to have remained fairly constant $1959 - 1989$, shortened to 1994, and then lengthened further to 2011. Figures of M3 for the US to 2006 have tended more towards the UK model.

By combining equations (5.13 and (5.14), a third expression for the entropy change for a polytropic process could be stated as:

$$ds = k\left(\omega\frac{dT}{T} + \frac{dV}{V}\right)_{rev} \tag{5.16}$$

Now for an isentropic condition, the expression in the brackets at equation (5.16) must be zero. Hence:

$$\omega\frac{dT}{T} = -\frac{dV}{V} \tag{5.17}$$

Figure 5.10 illustrates quarterly estimated values of money entropy change **ds** calculated from equation (5.16) for *assumed values* of value capacity coefficient ω for the UK economy of 0.65 and for the USA of 0.55, and by reference to annualised 4-quarter moving average percent rates of change in the index of trading value (velocity of circulation) **dT/T** and in output volume **dV/V**.

The assumed values of **ω** [UK 0.65, USA 0.55] were estimated by reference to the inverse of the mean values of the velocities of circulation for the UK (M4) of 1.52 and for the USA (M2) of 1.82, measured over the periods in the charts *[The mean figure of velocity of circulation for the USA on an M3 basis to 2006 calculates at 1.49, suggesting a value of **ω** on that basis of 0.67].* No account of any technical changes over time has been made, that might improve the results obtained.

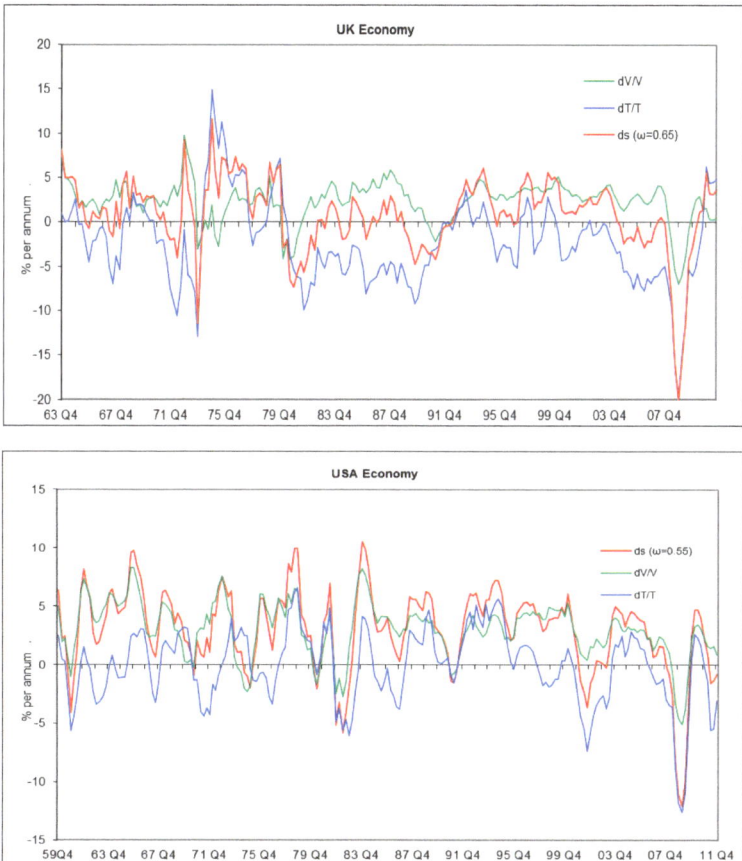

*Figure 5.10 Money entropy change **ds** for assumed values of value capacity coefficient **ω** and by reference to changes in output volume V and velocity of circulation **T**.*

It can be seen that money entropy change tends to go negative when volume change declines, and to increase when volume change goes up. It also tends to flow with change in velocity of circulation as that includes both volume and price changes. In addition, it will be noted that quite often when output volume change is positive then velocity change is negative. The two measures of money entropy change at equations (5.13) and (5.15) take this difference into account, in that the entropic index factors in front of dT/T and dV/V are different in construction. Readers will note, however, that mathematical manipulation of each of these factors, still produces the result for nil entropy change of $n = \gamma = (1+\omega)/\omega$.

From figure 5.10, if the two economies had been operating at near isentropic conditions, it might be expected that there would be little ebb and flow of ds about the zero line. Clearly this has not been the case, illustrated also by the significant variation in the elastic index n at figure 5.7. The fact that incremental entropy change ds tends to oscillate plus or minus either side of a minimum or zero level, however, suggests that the economic system endeavours to maximise or minimise entropy potential s in some fashion. This observation will be returned to later in this chapter.

By combining the charts at figures 5.7 for the elastic index, and the charts at figure 5.10 for the money entropy change, it will be noted that entropy change tends to go negative or downwards where the elastic index swings wildly. This effect appears also to occur when an increase in interest rates has occurred. Figure 5.11 sets out charts illustrating this effect.

Thus at equation (5.13) the entropic index is reduced to a low or even a negative value. The net effect on the money entropy change, however, depends upon whether changes in velocity of circulation are moving up or down, with a switchback occurring at about the zero point. Thus the position is complex. The charts of the elastic index at figure 5.11 show strategic points for both the UK and USA economies where the switchback effect occurred.

An explanation for this effect is given at figure 5.12, which sets out the locus of nil entropy change $n = \gamma = (1+\omega)/\omega$. It can be seen that large movements in the elastic index n occur if the curve of nil entropy gain is shifted to the left or to the right, effectively shortening or lengthening the apparent lifetime embedded in the value capacity coefficient. A further explanation, alluded to earlier in this chapter, is that volume change dV/V is small or going negative, whereby changes in the relationship $P_N V^n = C$ are magnified.

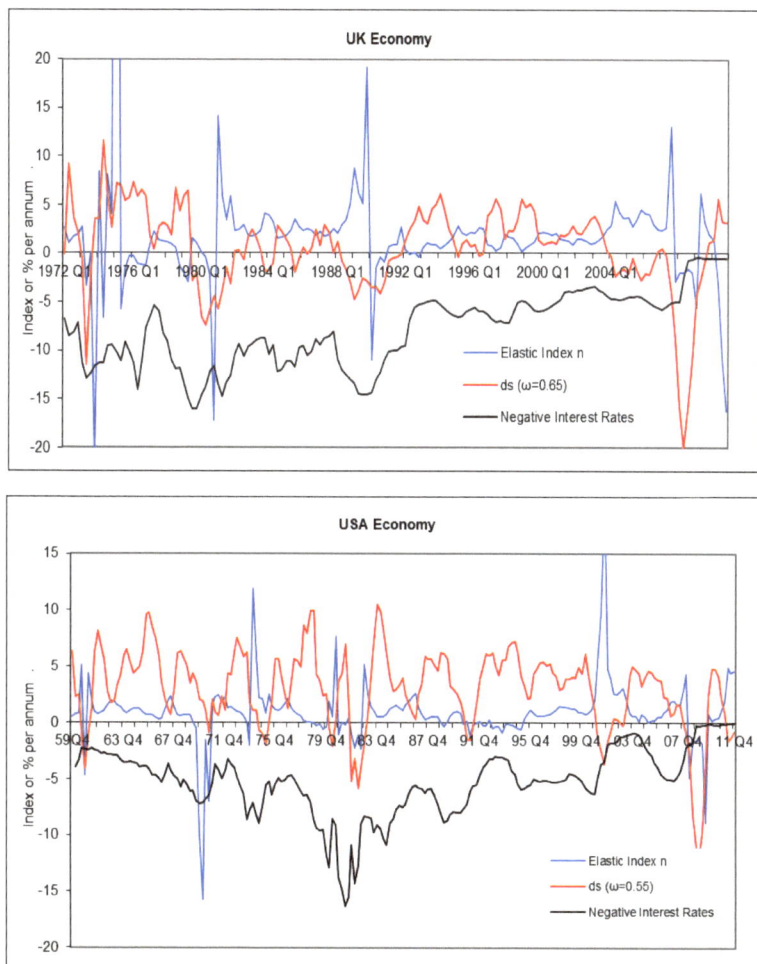

Figure 5.11 Elastic index, money entropy change and interest rates (3 month treasury rate). (Figures calculated from annualised 4-quarter moving averages, except interest rates which as actual)

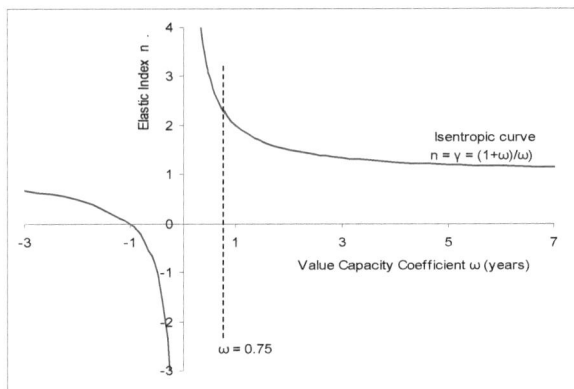

*Figure 5.12 Elastic Index **n** as a function of value capacity coefficient ω for nil entropy change*

5.3   Money Entropy and Interest Rates

To formulate the relationship of interest rates to the motivating force in an economy, derived in chapter 4 as entropy change, our starting point is to go back to the concept of *active* and *inactive* stocks, that culminated in a relationship between entropy change, output and constraints acting on output. A similar approach could be applied regarding the constraining forces acting on the demand for money balances as, in any transaction period, money continually flows out into the economy and back again, and in the opposite direction to output value flow.

It is generally accepted among economists that the demand for money is positively related to income and output, but negatively related to interest rates, and the standard textbook representation of demand for money is by reference to a curve of liquidity preference, a concept pioneered by Keynes (1936) in his book *'The General Theory of Employment, Interest and Money'*. Liquidity preference is effectively a curve of utility of money turned upside down, as in figure 5.13.

At this point it should be stated that we are not arguing in favour of either a Keynesian or Monetarist approach to economics, and whether management of the money supply or adjustments in fiscal spending via an IS/LM model [Investment-Saving / Liquidity preference-Money supply, Hicks, Hansen (1936)] provide the means to keep an economy in balance. We are, however, arguing that the thermodynamic analysis set out so far indicates that economies appear to operate with a polytropic relationship between specific price, output volume and velocity of circulation, with interrelating and continually changing elasticities. We are also arguing that there is a relationship between the concepts of utility and entropy, and therefore that the utility of money can be represented in terms of entropic value.

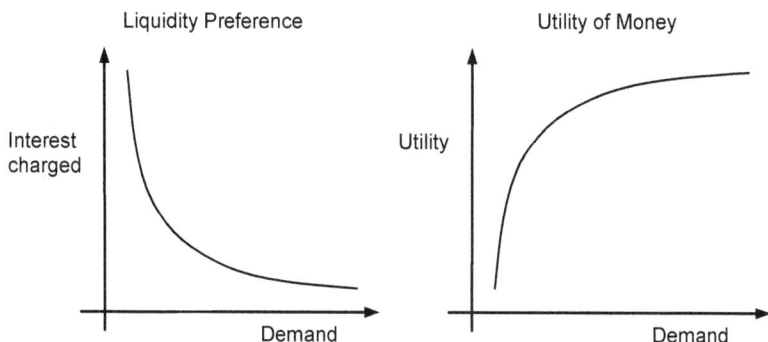

Figure 5.13 Liquidity preference and utility of money

Thus the inference of the analysis is that interest rates are negatively related to changes in money entropy value. The higher the level of money entropy change, and the higher price inflation, the more negative interest rates have to be to counteract the forces in the economy. Interest is therefore a form of value flow constraint and negative entropy change. The following relationship might therefore be posited between money balances **N**, interest rates **i**, the velocity of circulation **T** and entropy change **ds**:

$$\frac{dN}{N} = f\left(i, \frac{dT}{T}, ds_{money}\right) \tag{5.18}$$

With respect to interest rates, we imagine a cumulative interest index value I_t, which grows over time according to the level of interest rates. For money balances such an index will be composed primarily of short-term interest rates, typically a 3-month treasury rate or similar. Then for a constant interest rate **i**, the current cumulative index value I_t is satisfied over time by:

$$I_t = I_0 e^{it} \tag{5.19}$$

However, interest rates do vary over time according to economic conditions, and therefore our cumulative index of interest value I_t is calculated from a progression of variable interest rates i_t in each transaction period t (quarterly, annual) according to the formula:

$$I_t = I_0(1+i_1)(1+i_2)........(1+i_t)$$

$$= I_0 \prod_0^t (1+i_t)$$

where I_0 represents a convenient starting point.

Thence, for example, for adjacent points 1 and 2 in time: $I_2 = I_1(1+i_2)$, and $(I_2 - I_1) = i_2 I_1$, where i_2 is the interest rate applying in the specified year. Thus in general we could express the incremental rate of change of the index as dI/I, being equal to the variable interest rate i at any point in time.

Although money balances N exist primarily to facilitate flow of value within an economy, it is not unreasonable to accept that when they reside in a deposit account they will accumulate interest, as would money lent to borrowers accumulate chargeable interest. Such interest, when payable or chargeable, is included in the total of money balances. Thus there is likely to be a relationship between our cumulative interest index value I and both output value G and money balances N, depending upon the *active* use in economic output value, and the *inactive* time spent on balance (which is a function of the velocity of circulation T). It will be appreciated of course that with high inflation the index I is likely to rise significantly with larger interest rates implied. Likewise the number of money instruments N tends to rise as the currency is depreciated, and output value rises as prices and inflation escalate.

Figures 5.14a and 5.14b set out relationships between the cumulative interest index I (calculated from quarterly values of 3-month treasury rates), output value G (equal to price level P x output volume V), and nominal money balances N, according to the equation:

$$\left(\frac{I_2}{I_1}\right) = \left(\frac{G_2}{G_1}\right)^\theta = \left(\frac{N_2}{N_1}\right)^\tau \quad \text{or} \quad \frac{dI}{I} = \theta \frac{dG}{G} = \tau \frac{dN}{N} \quad (5.20)$$

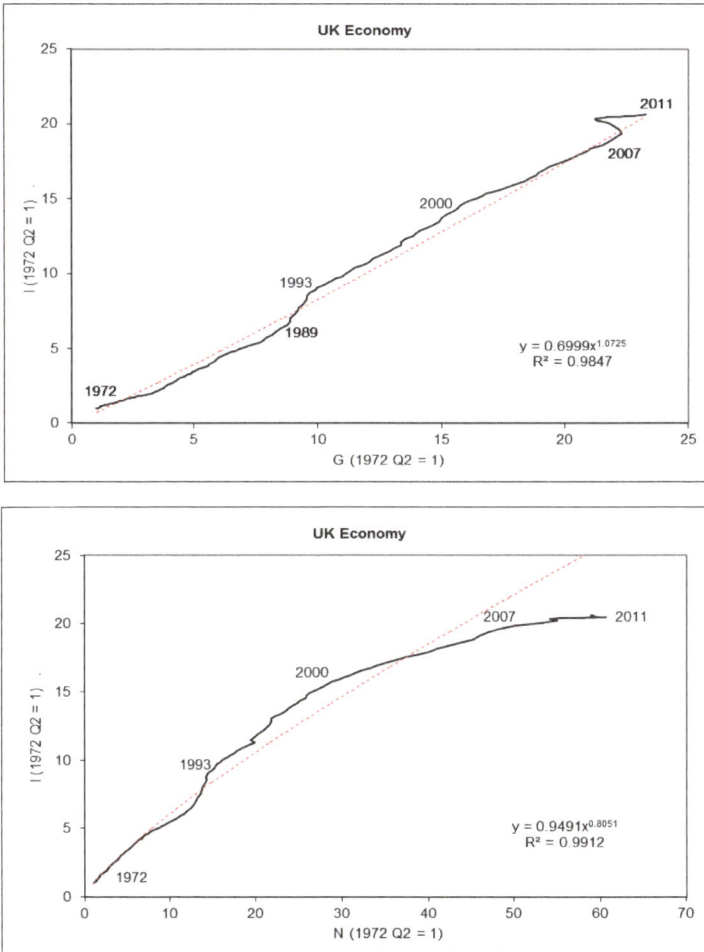

*Figure 5.14a Cumulative Money Interest Index **I** (3-month Treasury rate), Output Value **G** and Nominal Money Stock **N**. UK*

Where the elastic indices θ and τ can be variable, and where the third interrelationship is the velocity of circulation **T**, equal to **G/N**, shown earlier in this chapter. As with previous analyses, annualised 4-quarter moving averages were calculated. It was assumed that the cumulative index of interest value I_0 at the starting point of the time series for the each of the two economies was equal to 1.

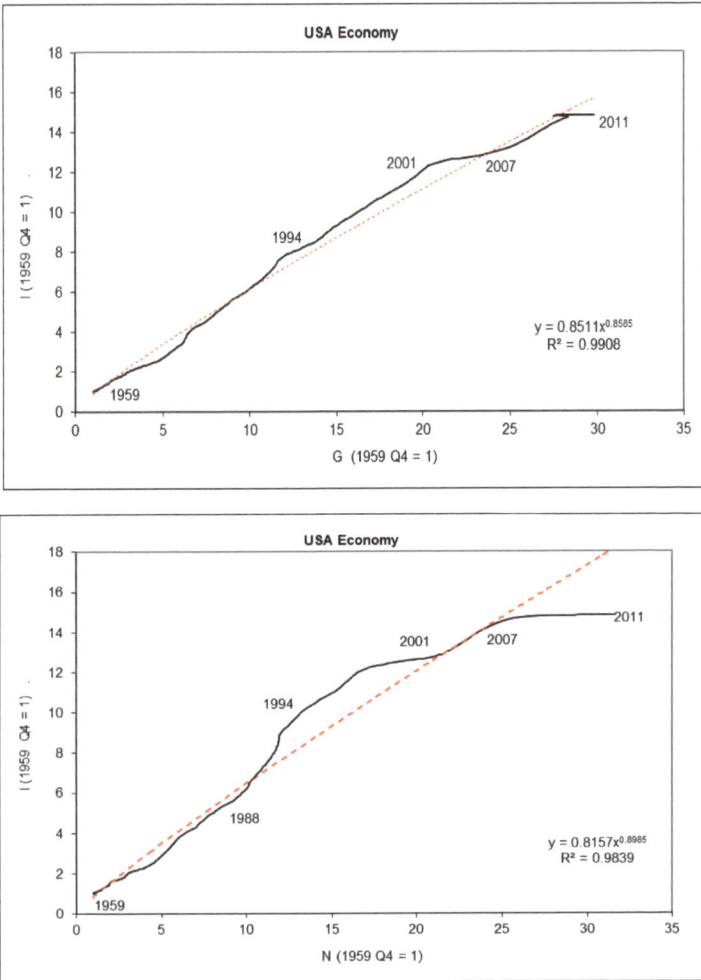

*Figure 5.14b Cumulative Money Interest Index **I** (3-month Treasury rate), Output Value **G** and Nominal Money Stock **N**. USA*

Though the regression coefficients of the relationships are quite high, there are significance differences in the slope of the curves over time, and hence the elasticities between the functions, as was the case also with the elastic index **n** between specific price and output volume, set out earlier in this chapter. Figure 5.15 sets out annualised 4-quarter moving averages of the factors to illustrate this point.

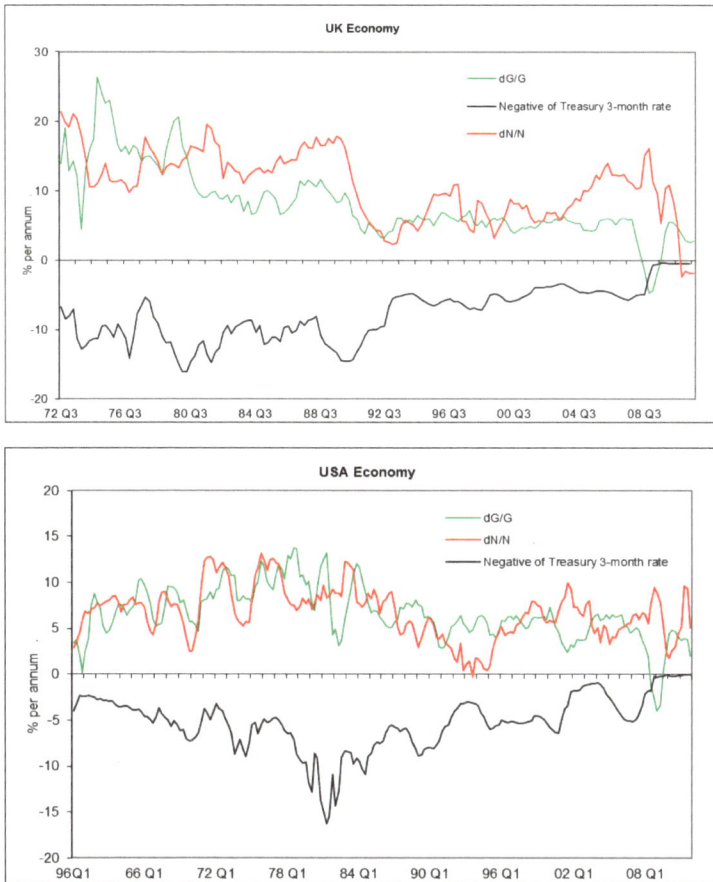

*Figure 5.15 Annualised changes (4-quarter moving averages) in output value **G**, nominal money stock N, and the negative of interest rates (3-month Treasury rate).*

On balance, it appears that the relationship of interest rates to change in output value is stronger than that to money balances.

Similar in construction to equation (4.25), chapter 4 for the production function, the following relationship is proposed to connect interest rates to both output **G** and entropy **s**:

*"In an economic system, the difference between the rate of change in output value flow **G** and the rate of change in the Index of Money Interest **I** is a function of changes in money entropy generated or consumed."*

This is expressed as:

$$ds_{money} = f\left[k_{money} \left(\frac{dG}{G} - \frac{dI}{I} \right) \right] \qquad (5.21)$$

where **dI/I** is the short-term interest rate **i** at any point in time. And **k** is the assumed 'productive content' of money.

The inference of equation (5.21) therefore is that an economic system will tend to operate with an entropic difference, positive when expanding and negative when contracting. A nil value would occur when output value rate of change is matched by interest rates. A highly positive value would equate to an economy seeking to maximise expansion, without being checked by interest rates. A negative value would equate to high interest rates coupled with low or negative output value growth.

Returning to our money analysis, if further we assume that the productive content **k** of equation (5.21) is equal to unity (£1, $1 etc), and the elastic factor θ at equation (5.20) is absorbed into the entropy change **ds,** then equation (5.21) becomes:

$$ds_{money} = \frac{dG}{G} - \frac{dI}{I} \qquad (5.22)$$

In this equation the rate of change in the cumulative interest index **I** is a function of the rate of change in output value **G**, but is negatively related to change in money entropy. Further, by substituting in the general money equation (5.3):

$$\frac{dG}{G} = \left(\frac{dP}{P} + \frac{dV}{V} \right) = \left(\frac{dN}{N} + \frac{dT}{T} \right)$$

we can derive a number of other relationships, of which the following is an example:

$$\frac{dN}{N} = \left(\frac{dI}{I} - \frac{dT}{T} + ds_{money} \right) \quad \text{or:}$$

$$\left(\frac{dN}{N} - \frac{dI}{I} \right) = \left(ds_{money} - \frac{dT}{T} \right) \tag{5.23}$$

It will be noted that this equation also has the same format as in our initial money hypothesis set out at equation (5.18). Thus the rate of change in money supply is equated to interest rates less the rate of change in the velocity of circulation plus the entropy change.

Figures 5.16 and 5.17 illustrate equations (5.22) and (5.23) for the UK and US economies:

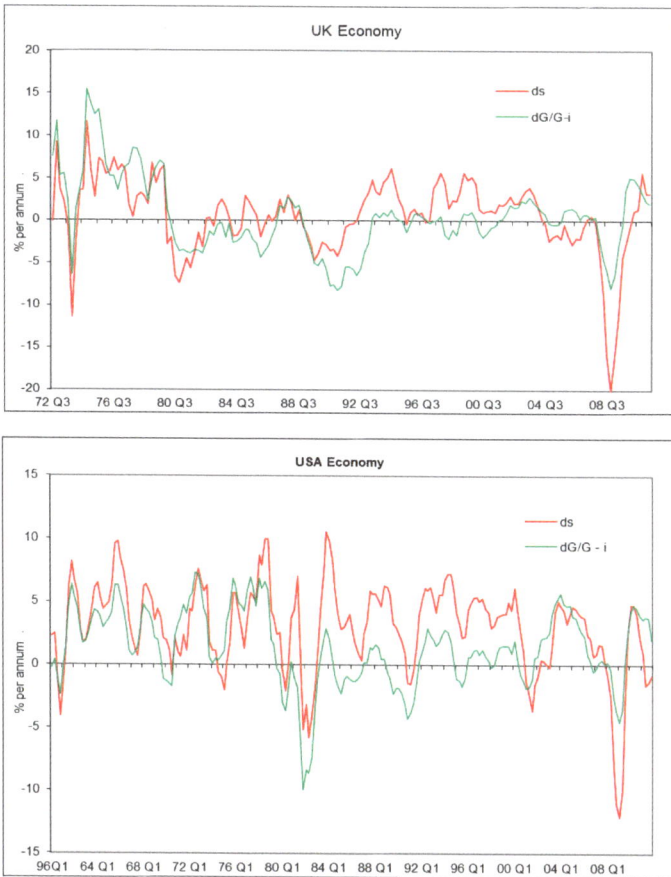

*Figure 5.16 Output value rate of chang **dG/G**, interest rates **i** and entropy change **ds**.*

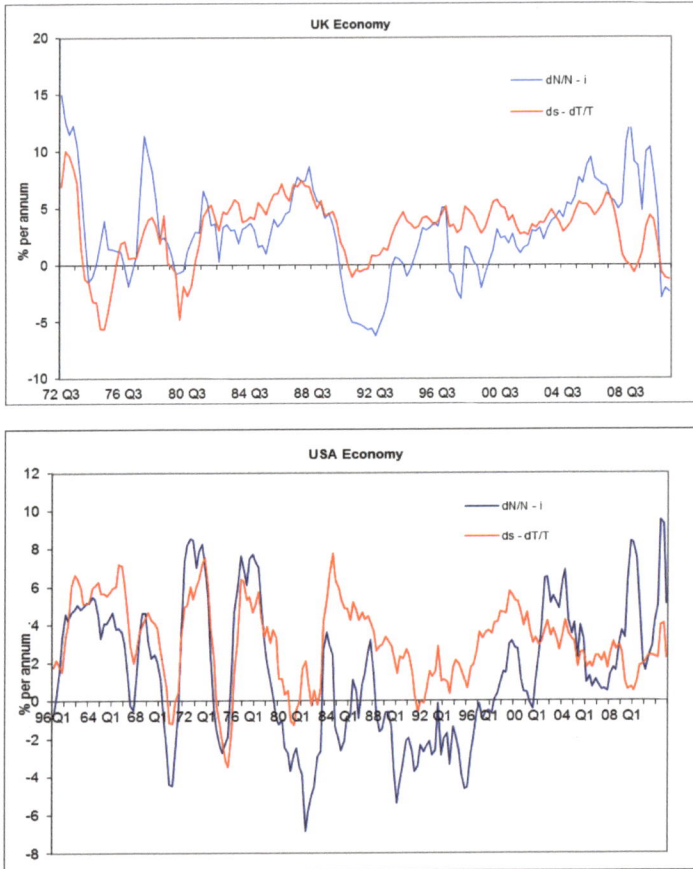

Figure 5.17 Money Balances N, interest rates, entropy and velocity of circulation

The relationships involving output value change **dG/G** appear to work better than with money balances, and this may have something to do with the way money supply has developed, veering away from a flat relationship with output value, with changes in velocity of circulation, as shown in figures 5.14a and 5.14b.

Because of the 'noise' inherent in the data, technical changes in the value capacity coefficient ω (and hence changes in long-run velocity of circulation), and changes in the impact of interest rates on an economy it is inevitable that there will significant deviations between values on either side of each equation. For example, the correlation coefficient for the UK economy of **ds** versus **dG/G-i** is of the order $R^2 = 0.44$, quite low. Thus much work is still required to improve and refine the relationships, perhaps modelling some of the technical changes in the velocity of circulation and other factors. The initial result is, nevertheless, quite interesting as each of the results appears to follow the ebb and flow of money entropy change. Figure 5.18 shows the differences per annum of entropy change **ds** versus (**dG/G − i**) for the UK and USA economies. Order-4 (UK) and order-5 (USA) trend lines have been drawn through each result.

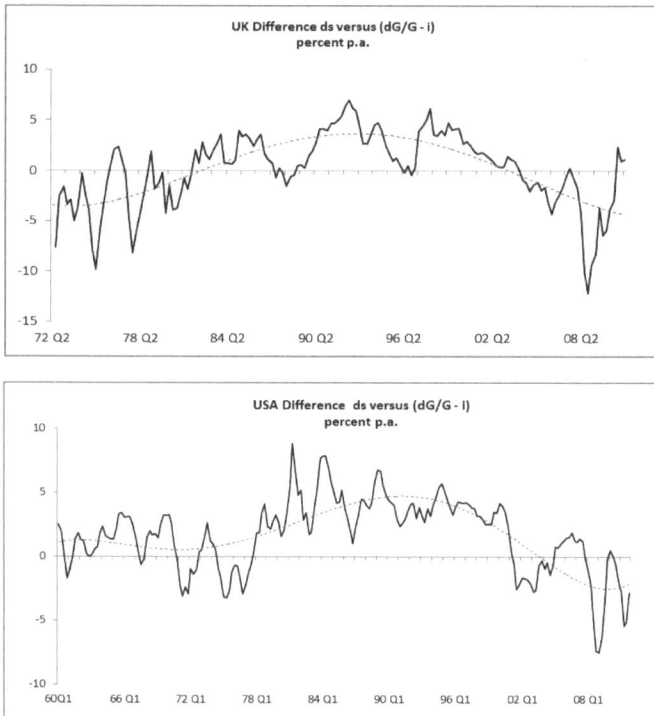

*Figure 5.18 Error Differences between entropy change **ds** and % change Output Value less interest rates (dG/G − i).*

The curves exhibit similar biases, and the trend lines for each have correlation coefficients of $R^2 = 0.54$ and 0.52 respectively. Thus correction for the biases would improve the correlation of the initial result significantly. Possible improvements may include better long-run modelling of the value capacity coefficient ω to estimate entropy change.

Accepting that there is a link between entropy change, output value and interest rates, the key indicator of portending change in an economy will be when sudden changes in the short-run elastic index **n** occur, as shown at figure 5.7.

From all the previous analysis and discussion, it will be appreciated that economic entropy change, positive or negative, represents a measure of whether an economy is likely to expand or contract. Thus knowing the conditions that define the direction is important.

It will be noted that the entropy change set out at equations (5.13) and (5.15) has two parts: a factor in the brackets called the entropic index, and either a change in velocity of circulation **dT/T** or a volume change **dV/V**. The factors are different for velocity and volume, though the solution for zero money entropy change is still the same **n=γ= (ω+1)/ω**.

$$ds_{money} = \left(\omega + \frac{1}{1-n} \right) \frac{dT}{T}_{rev}$$

$$ds_{money} = (\omega - \omega n + 1) \frac{dV}{V}_{rev} \qquad (5.24)$$

The two equations are linked by the elastic relationship:

$$\frac{dT}{T} = (1-n) \frac{dV}{V}$$

Thence, for zero money entropy change **n=γ= (ω+1)/ω** and we arrive as before at equation (5.17):

$$\omega \frac{dT}{T} = -\frac{dV}{V}$$

Thus the condition for positive money entropy change is therefore: Factor and multiplicand at equations (5.24) are either '*both* positive' or '*both* negative'.

And the condition for negative money entropy change is: Factor and multiplicand at equations (5.24) must be *opposite* in sign i.e. factor positive and multiplicand negative, and vice versa.

Figure 5.19 sets out the loci of the entropic indices at equation (5.24) for varying values of elastic index **n** and value capacity coefficient ω.

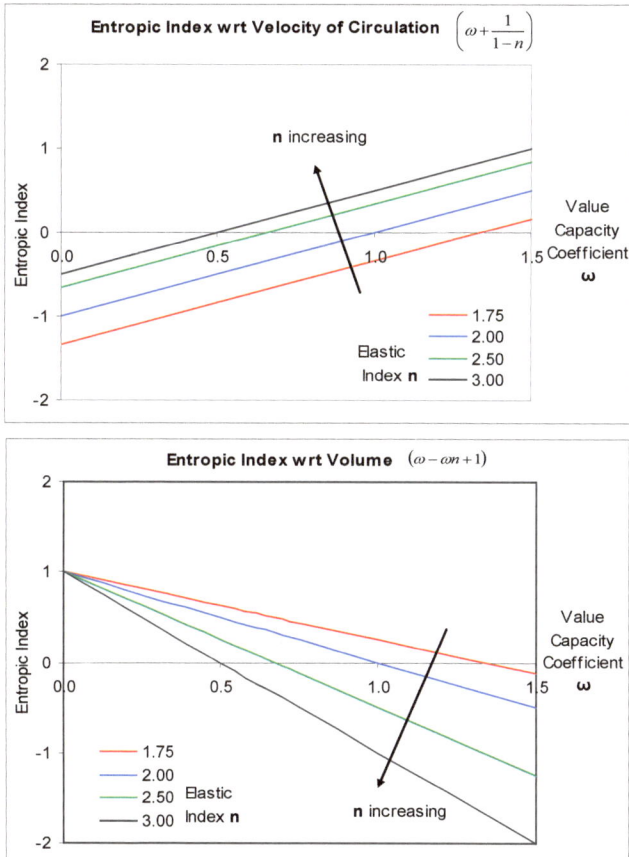

Figure 5.19 Loci of Entropic Indices

CHAPTER 6   LABOUR AND UNEMPLOYMENT

Having set out the general principles of thermodynamics as applied to
economic processes, and seen how money interacts thermodynamically with
the economy, it is of interest to return to analysis of the wage sector. It will
be recalled that in chapter 2 an expression for the total labour stock process
was developed at equation (2.12):

$$P_L v_L = w_E = k_M T_E \tag{6.1}$$

In this equation, P_L is the price of a volume throughput of labour stock and
v_L is the specific volume rate. v_L essentially represents the proportion of a
lifetime of a labour output that is spent each year by a single member of the
labour force. Thus, for a person with a 45-year lifetime in the labour force,
v_L will be $1/45^{th}$ per annum of the total potential output volume contribution
that such as person can provide. Clearly we are talking mean values or
averages here, but that is the essence of the thing at a macro level.

In the equation, T_E is an overall index of trading value with respect to
human labour stock; reflecting not only the throughput of human-power and
relative prices, but the accumulated value from resources and other inputs.
T_E is not just a velocity of circulation with respect to labour therefore, but is
a measure of the overall value of output per capita of an economic system
that humankind attributes to itself. Since k_M for money is equal to 1 (£1, $1
etc), T_E might therefore be represented by a wage rate w_E which would be a
mean rate.

Given that wages account for most of output in standard economic
accounting, one can imagine that the labour sector is deemed to have a
similar central function as money does. A general polytropic process can
therefore be posited to represent the relationship between the price P_L of
labour and specific volume rate v_L. It will be recalled from equation (3.46)
chapter 3 that this is of the form:

$$P_L \left(v_L \right)^n = Z \tag{6.2}$$

Where Z is a constant and n is an *elastic* index. And by taking logs and
differentiating we have:

$$\frac{dP_L}{P_L} = -n\left(\frac{dv_L}{v_L}\right) \tag{6.3}$$

High values of the elastic index are associated with an inelastic position, whereby output volume is not affected much by changes in price; and vice-versa.

Proceeding still further, an expression could be constructed for the incremental entropy change for a polytropic process (as was set out in chapter 5 for the money system), but this time applied to the labour sector. As per equation (3.52) in chapter 3, this will have the form:

$$ds_{labour} = k\left(\omega + \frac{1}{1-n}\right)\frac{dT_E}{T_E}$$

Where **k=1** (£1, $1 etc) and ω is the value capacity coefficient – effectively equal to the lifetime of our person in the labour force. In the equation **n** is the elastic index as per equation (6.3) and T_E in the index of trading value. However since in a wage system the index of trading value T_E is equivalent to the wage rate w_E, we can write:

$$ds_{labour} = k\left(\omega + \frac{1}{1-n}\right)\frac{dw_E}{w_E} \tag{6.4}$$

Further, from equation (6.4) for incremental entropy change, we can also imagine an isentropic position, whereby entropy change is zero. And from the analysis at chapter 3 this condition is satisfied is when $n = (\omega + 1)/\omega$. But whereas in a money system the value capacity coefficient ω is in the range 0.5 – 1 year (UK and USA values – see chapter 5), for the labour sector this is likely to be much longer. Figure 6.1 sets out a chart for the isentropic condition to a span of 50 years.

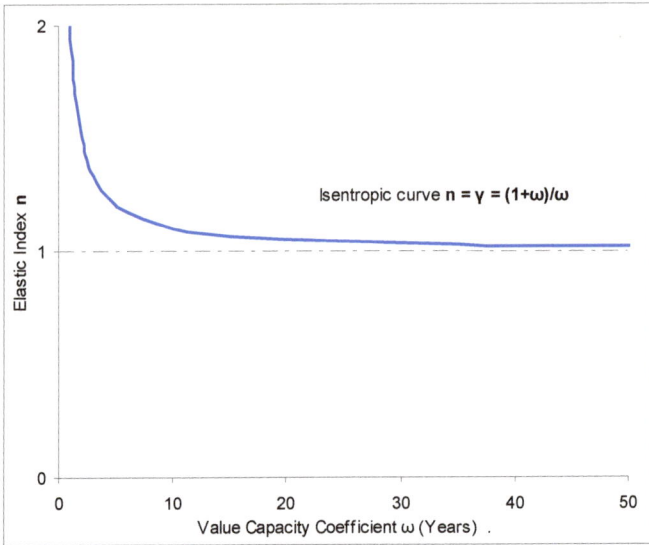

Figure 6.1 Elastic Index as a function of Value Capacity Coefficient

In general the elastic index **n** is not much in advance of 1, signifying that changes in price have a significant effect on volume output and, conversely, a change in output volume can also affect significantly the price of labour.

We now turn to the subject of unemployment and its impact on economic output. It is commonly accepted that the level of wage rates is affected by the rate of unemployment, of which the Phillips curve [William Phillips (1958)] (see figure 6.2) is the most well-known means of illustrating this effect; though there are other more modern variants, including NAIRU (non-accelerating inflation rate of unemployment). When unemployment is high, wage growth tends to be low, and when unemployment is low wage rates tend to rise. This is not always the case, as other factors can impact on demand. History has shown that the Phillips curve tends to shift. In the 1970's, many countries experienced high levels of both inflation and unemployment – stagflation, which changed the relationship of the curve.

It is logical to deduce that a rise in unemployment not just takes out a proportion of the workforce, but that it also takes out a proportion of output value flow that might otherwise have been generated, since at that point output value in the economy can no longer support the additional costs. In a similar manner to our money system at chapter 5, the following hypothesis might be proposed:

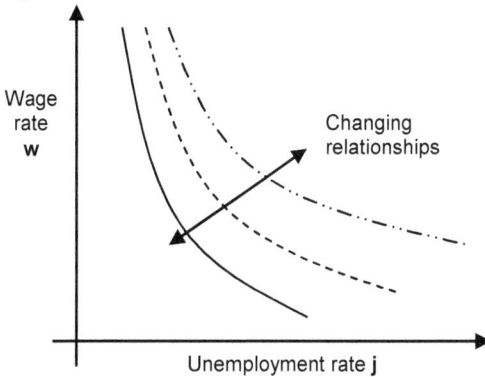

Figure 6.2 Phillips Curve

"In an economic system, the difference between the rate of change in output value attributed to the labour sector and the rate of change in lost value from unemployment is a function of the change in labour entropy generated or consumed."

This could be set out as:

$$ds = f\left[k_{money}\left(\frac{dG}{G} - \frac{dJ}{J} \right) \right] \qquad (6.5)$$

Where **G** represents output value attributed to the labour sector that the economy can support, and **J** is the potential output value flow loss taken out through unemployment, which the economy cannot support.

To test this hypothesis, we make a number of substitutions to derive the structural relationship between output value flow and lost output value.

First, from equation (2.8) chapter 2 we could substitute wage rate w_E for the economy *[GDP per capita]* multiplied by labour stock in employment N_L for output value *[it is accepted that this is a simplification involving also profits]*:

$$G = w_E N_L \qquad or \qquad \left(\frac{dG}{G} \right) = \frac{dw_E}{w_E} + \frac{dN_L}{N_L} \qquad (6.6)$$

Second, we could set out the lost output value **J** as:

$$J = w_E j N_T \quad \text{or} \quad \left(\frac{dJ}{J}\right) = \frac{dw_E}{w_E} + \frac{dj}{j} + \frac{dN_T}{N_T} \tag{6.7}$$

where w_E is the wage rate, **j** is the rate of unemployment, and N_T is the total labour stock.

Last we have that the employed labour force N_L equals total labour stock N_T, less the unemployed jN_T. Thus by substitution, taking logs and differentiating, we can write:

$$N_L = N_T(1 - j) \quad \text{or} \quad \left(\frac{dN_L}{N_L}\right) = \frac{dN_T}{N_T} - \frac{dj}{1-j} \tag{6.8}$$

We now assume that the functional relationship at equation (6.5) is unity and that k_{money} is equal to 1, as we did for the chapter on money. Thus:

$$ds_{labour} = \left(\frac{dG}{G}\right) - \left(\frac{dJ}{J}\right) \tag{6.9}$$

Substituting equations (6.6) - (6.8) back into equation (6.9) we have:

$$ds_{labour} = \left(\frac{dw_E}{w_E} + \frac{dN_T}{N_T} - \frac{dj}{1-j}\right) - \left(\frac{dw_E}{w_E} + \frac{dj}{j} + \frac{dN_T}{N_T}\right)$$

Hence:

$$ds_{labour} = -\left(\frac{dj}{j} + \frac{dj}{1-j}\right)$$

Or:

$$(s_2 - s_1)_{labour} = \ln\left(\frac{j_1}{j_2}\right)\left(\frac{1-j_1}{1-j_2}\right) \tag{6.10}$$

which expresses entropy change as a function of the unemployment rate.

It will be noted that since labour stock **N** has cancelled out, our entropy change is confirmed as per unit of labour stock. Further, given that **j** is generally small in relationship to 1, this expression reduces approximately to:

$$ds_{labour} \approx -\left(\frac{dj}{j}\right)$$

Or:

$$(s_2 - s_1)_{labour} \approx \ln\left(\frac{j_1}{j_2}\right) \qquad (6.11)$$

which expresses incremental entropy change of the labour sector as a function of the negative of the rate of change in the unemployment rate. Thus if entropy change is positive the rate of change in the unemployment rate is likely to be negative, and if entropy change is negative the rate of change in the unemployment rate is likely to be positive.

The key to the process is the entropy change. Entropy gain provides the impetus to drive an economic system forwards, and a decline in entropy will drive it backwards into recession or deflation.

Given that in traditional economic parlance wages are deemed to form the main part of costs, it is not unreasonable to deduce that labour entropy change will follow money entropy change to a first degree of approximation. At the isentropic point, for example, where

$$n = \left(\frac{\omega+1}{\omega}\right)$$

then nil entropy change can be satisfied by a labour value capacity coefficient ω of 50 combined with an elastic index n of 1.02, and similarly, by a money value capacity coefficient of $\omega = 0.75$ combined with an elastic index n of 2.333. The only element of variation is likely to be either side of the isentropic position.

By way of illustration, the charts at figure 6.3 set out a comparison of *money entropy* change Δs, taken from the money derivations in chapter 5, versus the negative percent rate of change in unemployment rate dj/j, as per equation (6.11).

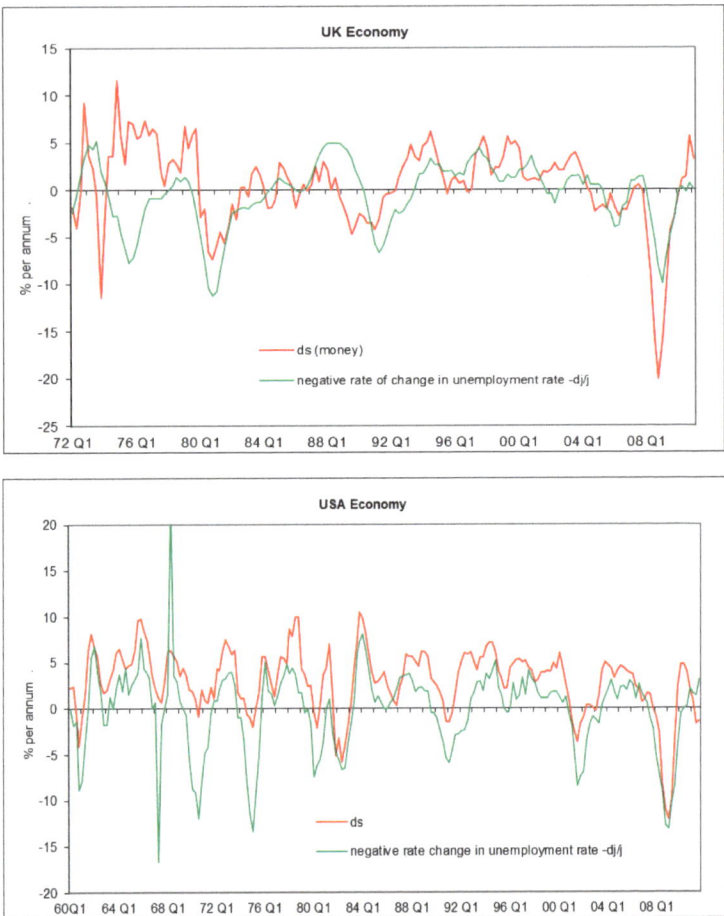

Figure 6.3 Changes in entropy, and the negative of rate of change in unemployment rates

Data of unemployment rates was taken from US Bureau of Labor Statistics (www.bls.gov) and Economic Trends Annual Supplement (www.statistics.gov.uk). Quarterly changes in the unemployment rate were calculated from 4-quarter moving averages of the rates.

The data presented appear to indicate that negative rates of change in unemployment rates in the USA follow the ebb and flow of changes in money entropy. The evidence is less conclusive in the case of the UK, however, perhaps owing to the large inflationary period in the 1970s.

Finally, we could combine equation (6.10) with equation (6.4) to arrive at an equation setting out the rate of change in the wage rate in relation to the rate of change in the unemployment rate:

$$ds_{labour} = \left(\omega + \frac{1}{1-n}\right)\frac{dw_E}{w_E} = -\left(\frac{dj}{j} + \frac{dj}{1-j}\right)$$

Or:

$$\left(\frac{dj}{j} + \frac{dj}{1-j}\right) = -\left(\omega + \frac{1}{1-n}\right)\frac{dw_E}{w_E} \qquad (6.12)$$

Since the rate of unemployment **j** is small, this can be reduced to:

$$\left(\frac{dj}{j}\right) \approx -\left(\omega + \frac{1}{1-n}\right)\frac{dw_E}{w_E} \qquad (6.13)$$

This equation describes the rate of change in the unemployment rate **dj/j** as being negatively related to the rate of change in wage rates **dw_E/w_E**, moderated by a function of the elastic index **n** and the value capacity coefficient **ω**. This result follows the format of the inverse relationship of the Phillip curve set out in figure 6.2, with the main influencing factor affecting the relationship being the elastic index **n**, which has the effect of moving the curve in the manner shown. Empirical analysis is required to take this aspect further, which is outside the scope of this book.

It should be emphasised again that money and labour entropy generation in an economic system act at the margin, in regulating output value flow, but do not affect the main body of entropy generation arising from consumption of resources and products.

CHAPTER 7   INVESTMENT AND ECONOMIC ENTROPY

At chapter 4 a discussion was set out concerning the principle of maximising entropy gain, by which decision makers seek to maximise benefits to their businesses and organisations. This principle relates also to that of utility, discussed at chapter 3, by which consumers endeavour to maximise benefits to themselves from among a range of opportunities on which to spend a limited income. In this chapter we outline the way in which this principle interacts with investment decision making, in particular, discounted cash flow, bonds and yield.

It will be recalled from equation (4.25) that entropy change was related to rate of change in active output value less the rate of change in any impacting constraint toward inactiveness, all provided that rates of change in prices were equalised between output and constraint.

$$ds = k_{money} \left[\frac{dG_{Oa}}{G_{Oa}} - \frac{dG_{Oc}}{G_{Oc}} \right] \quad \textit{[subject to } \mathbf{dP/P} \textit{ active} = \mathbf{dP/P} \textit{ inactive]}$$

And in terms of money this equation became that stated at equation (5.22):

$$ds_{money} = \frac{dG}{G} - \frac{dI}{I} \tag{7.1}$$

Where **ds** was entropy change, **dG/G** was the rate of change in output value and **I** was a cumulative index of investment value that increased each period/year by the prevailing interest rate **i** at any time, which can go up and down. Thus **dI/I = i**. Further, turning the equation round, one could project:

$$\frac{dG}{G} = \frac{dI}{I} + ds_{money} \tag{7.2}$$

This equation expresses output potential as a function of the current rate of interest, plus an entropy change indicating the utility or expectations that humankind has concerning forward economic conditions; positive when future returns might be expected to exceed current rates of interest, and negative when future returns might be expected be lower than current rates of interest. Thus the equation sets out expectations with respect to the climate for investment.

7.1 Project Investment and Discounted Cash Flow

In respect of project investments, a generalised form of investment return is of the form:

$$surplus / deficit = -N_P + \lfloor A_1 + A_2 + ... A_\xi \rfloor \qquad (7.3)$$

Where N_P is the initial investment project outlay and A_1, A_2 .... is a stream profits/payments received over a period of ξ years. Because such investments and projects take place into the future, it is common practice to discount each future receipt back to the time of the initial outlay using an appropriate rate of return r. Thus is born the familiar equation for discounted cash flow (DCF) involving a Net Present Value (NPV):

$$NPV = -N_P + \frac{A_1}{(1+r)^1} + \frac{A_2}{(1+r)^2} + ... + \frac{A_\xi}{(1+r)^\xi} \qquad (7.4)$$

Or in shorthand:

$$NPV = -N_P + \sum_{x=1}^{\xi} \left(\frac{A_x}{(1+r)^x} \right) \qquad (7.5)$$

Hence if the net present value is positive then the project is deemed to have a chance of beating prevailing investment returns.

An alternative method of presenting the Net Present Value is by what is known as the 'Internal Rate of Return' or IRR, which might represent the minimum rate of return r required, such that the Net Present Value becomes zero. Thus:

$$N_P = \sum_{x=1}^{\xi} \left(\frac{A_x}{(1+r)^x} \right) \qquad (7.6)$$

Where r is the Internal Rate of Return IRR, which could be compared to prevailing investment returns to assess the profitability of the project / investment.

When comparing several projects with each other, however, it is generally recognised that the NPV form of discounted cash flow analysis at equation (7.5) is to be preferred over IRR.

With perfect knowledge, both of future rates of return and of each project return amount A_x, equation (7.5) could be modified as in equation (7.7):

$$NPV = -N_0 + \sum_{x=1}^{\xi} \left(\frac{A_x}{(1+r_x)^x} \right)$$ (7.7)

Where r_x is the rate of return in each period into the future up to period ξ.

Quite evidently, however, managers and decision makers are not blessed with perfect knowledge of what the future may hold, and experience suggests that it is prudent to set a project rate of return in excess of prevailing rates in order to allow for risk and uncertainty.

Figures 5.7 and 5.16 at chapter 5, for example, show several periods of economic uncertainty in the UK and USA, when the elastic index n was very high/low, and when economic entropy change Δs became negative. At such times there is additional risk attached to projects concerning the future, and a short term view might be preferred to a long term one.

A possible means of measuring and incorporating future uncertainty into the discount equation, therefore, is to modify the rate of return to include an entropy function. Thus at any point in time ($t = x$) in the future we could write:

$$r_e = r_x + \Delta s_x$$ (7.8)

Where r_e is the expected rate of return.

However, given that knowledge of the future is unknown, in particular as to whether Δs_x is positive or negative in any year, decision makers have to base their assessment on the present. Suppose we set out two equations for equation (7.8), one for future expectations and one for the present, viz:

$$r_e = r_x + \Delta s_x$$
$$r_e = r_0 + \Delta s_0$$

Then by subtracting one from the other and assuming r_e is the same for each, we have:

$$r_x = r_o + \left(\Delta s_0 - \Delta s_x \right) \qquad \text{or}$$

$$r_x = r_o + \Delta s_0 \left(1 - \frac{\Delta s_x}{\Delta s_0} \right) \tag{7.9}$$

Thus the future rate of return at time $t = x$ could be expressed as the present rate of return at time $t = 0$, plus the present entropy change modified by an entropy difference function.

It will be recalled from chapter 5 of this book that entropy change was linked to utility, and that other researchers also have highlighted similarities between the two concepts. In essence therefore, what is required for the factor in the brackets at equation (7.9) is a discounted utility or *entropy* function that enables decision makers to put a weight on future outcomes, given that the further ahead one attempts to look, the less will be known. Indeed that is exactly what occurs in the process of discounted cash flow and in discounting bond coupon payments to present values. For the purposes of this analysis we might posit an *entropy decay* function of an exponential form though, as highlighted later in the section on bonds, others might also be considered. Thus for the moment we assume:

$$\Delta s_x = \Delta s_0 e^{-\varepsilon x} \tag{7.10}$$

Where ε is a decay factor.

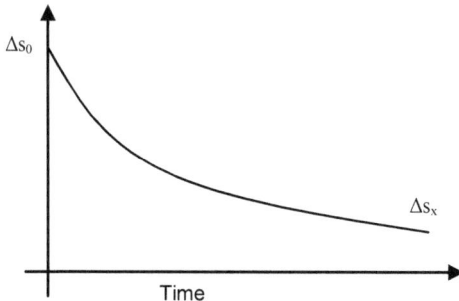

Figure 7.1 Exponential entropy decay function

Thus future changes in entropy are estimated as being equal to the current one modified by a decay function, expressing a view as to the extent to which the current position could be projected forward. Thus from equations (7.8) and (7.9) the expected rate of return at any point could be expressed as:

$$r_e = r_0 + \Delta s_0 \left(1 - e^{-\varepsilon x}\right) \tag{7.11}$$

Where the decay factor ε is set by decision makers; having regard for current position and for the time horizon of a particular project. Clearly at times of low expectations with a negative entropy change, the expected return might be below current interest rates, and investment for the future is not promoted, and vice-versa for times of high expectations. A worked example in support of this approach will be set out later in this chapter in respect of bonds and gilt-edged stocks.

7.2 Annuities

The format of an annuity is similar in structure to that for project analysis, with the difference that all of the period payments **A** are the same.

Thus, assuming a single rate of return, the equation for the net present value is reduced to:

$$NPV = \frac{A}{(1+r)^1} + \frac{A}{(1+r)^2} + ... + \frac{A}{(1+r)^\xi} \tag{7.12}$$

Or in shorthand:

$$NPV = \sum_{x=1}^{\xi} \left(\frac{A}{(1+r)^x}\right) \tag{7.13}$$

Mathematical manipulation reduces this to:

$$NPV = \frac{A}{r}\left[1 - \frac{1}{(1+r)^\xi}\right] \tag{7.14}$$

Or in exponential format:

$$NPV = \frac{A}{r}\left[1 - e^{-r\varsigma}\right] \tag{7.15}$$

In respect of risk and uncertainty, a similar argument can be applied to an annuity as to that for a project investment, concerning the expected rate of return, in the knowledge of changes in entropy affecting matters and, similar to equations (7.8) and (7.11), we could posit an expected rate of return of the form:

$$r_e = r_\xi + \Delta s_\xi$$
$$r_e = r_0 + \Delta s_0 \left(1 - e^{-\varepsilon\xi}\right) \tag{7.16}$$

Thus an annuity with negative entropy expectations would not likely be attractive to an investor, unless a very high yield was involved.

7.3   Bonds and Gilt-Edged Securities

Bonds and gilt-edged securities, issued by governments and corporations, are structured around two main types: conventional bonds (the majority) involving fixed coupon income and redemption values, and index-linked bonds involving index-linking of both coupon and redemption values in line with a consumer price index, to provide for an element of inflation-proofing.

A conventional fixed-interest bond cash-flow construction is of a similar form to that for project investments and annuities, but involving also a return of the original investment to the lender (at par) at the end of the investment period. As with an annuity, fixed coupon payments **A** (annual, half yearly) are paid to the lender between the beginning and end of the investment period, but the coupon payment **A** here is defined as a coupon rate **c** multiplied by the par value **Φ** returnable at the end of the investment period ξ, that is **A = cΦ**. The coupon rate **c** is generally set with due regard for rates of return and economic conditions prevailing at the bond issue date. Thus in cash-flow terms for *annual* coupons for a bond investment period of ξ years with an issue price of **N_P** we have:

$$(\xi c + 1)\Phi - N_P = surplus\,/\,deficit \tag{7.17}$$

The surplus/deficit from investing in a bond is generally absorbed by reformatting the equation in terms of a discount interest rate or yield to redemption **r**, with the purchase price N_P effectively balancing out income and par value against prevailing interest rates, as follows:

$$N_P = \frac{c\Phi}{(1+r)^1} + \frac{c\Phi}{(1+r)^2} + \dots\dots + \frac{c\Phi}{(1+r)^\xi} + \frac{\Phi}{(1+r)^\xi} \qquad (7.18)$$

Thus a bond price N_P above par value Φ, would imply a net income stream above prevailing interest rates, and vice-versa, a bond price below par value would imply a net income stream below prevailing interest rates. By definition for one particular bond, all components are fixed except the price N_P, the period ξ remaining to maturity and the yield to redemption **r**. Equations (7.19) set out shortened versions of equation (7.18) using standard or exponential notation:

$$\frac{N_P}{\Phi} = \left[\frac{1}{(1+r)^\xi} + \frac{c}{r}\left(1 - \frac{1}{(1+r)^\xi}\right) \right]$$

$$\frac{N_P}{\Phi} = \left[e^{-r\xi} + \frac{c}{r}\left(1 - e^{-r\xi}\right) \right] \qquad (7.19)$$

By way of example, the charts at figure 7.2 illustrate the relationship between the three factors: yield to redemption **r**, bond price N_P and maturity period ξ, for a coupon rate **c** = 3%. In each case one variable is represented as an isopleth, linking points with the same value.

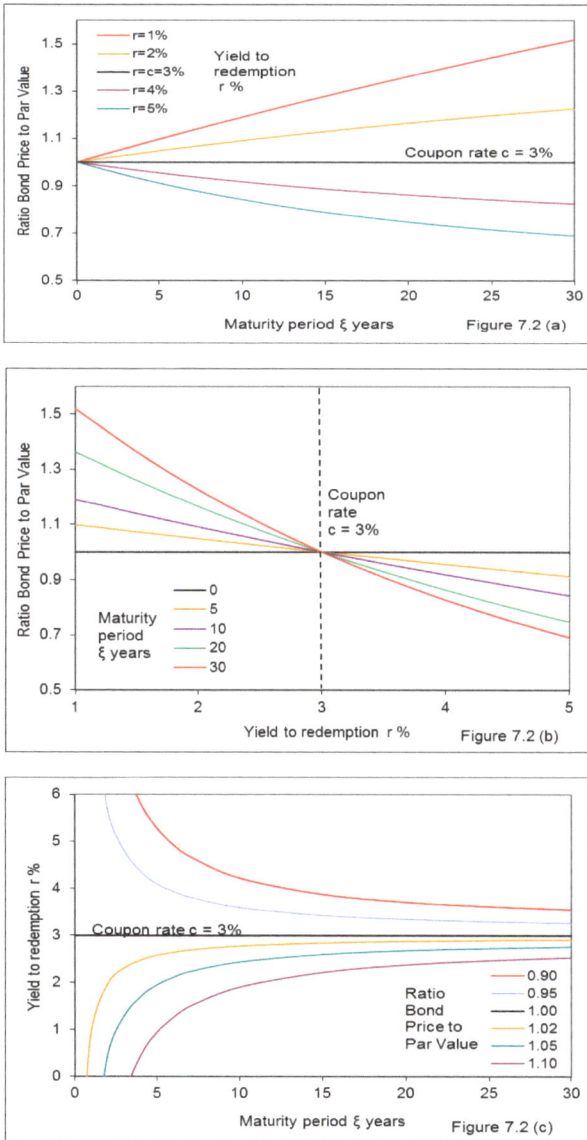

Figure 7.2(a) – (c) Conventional Bonds/Gilts. Relationships between yield to redemption, maturity period and the ratio of bond price to par value.

From equation (7.19) the current yield **y** of a bond is expressed as:

$$y = \frac{c\Phi}{N_P} \qquad (7.20)$$

And the solution $\xi = \infty$ years to maturity (an undated stock) gives:

$$r = y = \frac{c\Phi}{N_P} \qquad (7.21)$$

The solution $\xi = 0$ years to maturity gives $N_P = \Phi$, and the solution $N_P = \Phi$ gives $r = c$ for all ξ.

Yield to redemption r is generally regarded as more meaningful than current yield, and forms a basis to compare bonds in the same class and credit quality. Index-linked bonds generally have lower yields, combined with higher purchase prices, compared to standard bonds. A yield curve can be constructed to link bonds with different maturities and coupon rates. Yield curves are mostly upward sloping as maturity lengthens, reflecting future expectations and a risk premium for holding long-term securities.

It is not always the case, however, that long-term yields exceed short term rates. On occasion, the reverse can occur, with long-term yields being below short-term rates, producing a negative spread. A negative spread is often regarded as portending an increased risk of impending recession, with investors marking down their view of the future with respect to the present.

Figures 7.3 for example, illustrates some more recent changes in terms of British conventional gilt-edged stocks. In 2007 long-term yields were at a discount to short-term yields. There followed a virulent recession, but by 2010 the yield spread had once more resumed a positive position.

Referring to equation (7.19), changes in the shape of the curves occur when different prices N_ξ of bonds are attached across the maturity periods, and coupon rates, such that short-term yields rise or decline with respect to long-term yields.

The three widely followed theories explaining the curvature of the yield curve are 'Pure Expectations', 'Liquidity Preference' and 'Preferred Habitat'. The first theory reflects investors' expectations of future interest rates. The second theory adds in a premium for long term rates to compensate for added risk of money being tied up for a longer period, and the third theory assumes that investors have distinct investment time horizons and require a premium to buy outside their preferred maturity. Estrella & Trubin (2006) offer some guidelines on constructing yield curve indicators with regard to predicting future economic activity.

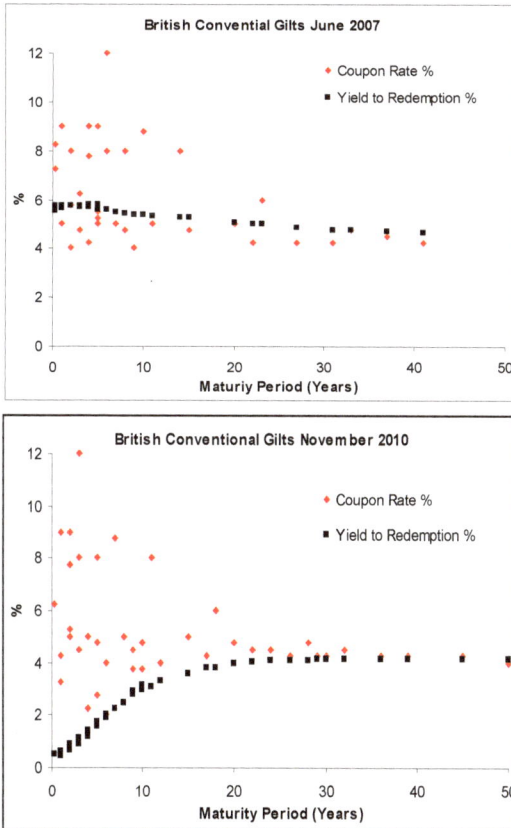

Figure 7.3 Yield to redemption and Coupon rates British Funds

Figure 7.4 sets out historical charts of interest yields of long and short-dated stocks in the UK and USA. Historically, yields of long-term funds in the USA have been mostly in excess of those of short term funds, reflecting a positive yield spread, but in the UK there has been a more balanced variation.

170

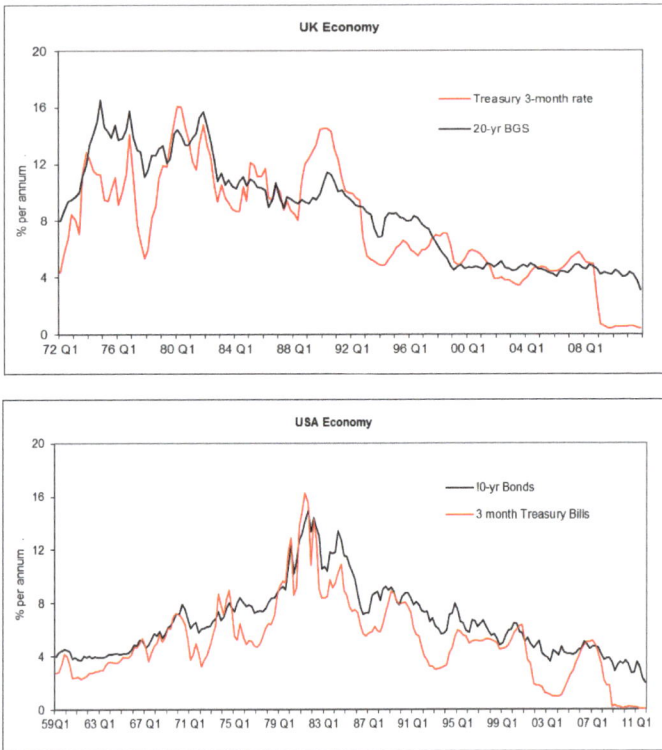

Figure 7.4 20-year BGS for the UK and 10-year bond rate for the USA, set against 3-month Treasury Rate

A significant body of empirical research exists, directed to estimate yield curves, based on real data of yields obtained. Chief among these are Nelson & Siegel (1987), Svensson (1994), Waggoner (1997), Anderson & Sleath (1999) and Bacon (2004). Nelson & Siegel's model of the instantaneous forward rate curve is of the form:

$$f(m,\beta) = \beta_0 + \beta_1 e^{-(m/\tau)} + \beta_2\left(\frac{m}{\tau}\right)e^{-(m/\tau)} \qquad (7.22)$$

Where β_0, β_1 and β_2 are long-run, short-run and medium term interest rate components, τ is a decay component and **m** is the maturity at which the forward rate is evaluated. Such equations do not, however, altogether explain the forces impacting on yields, only how yields of bonds having varying coupon rates and maturity dates fit with each other.

We now refer again to our thermo-economic analysis, and equation (7.2) linking output potential to current interest rate return **i** (= **dI/I**), and money entropy change **ds**.

$$\frac{dG}{G} = \frac{dI}{I} + ds_{money}$$

Turning this equation round we could write:

$$\frac{dI}{I} = \frac{dG}{G} - ds_{money} \qquad (7.23)$$

In respect of a bond, it was noted earlier that the coupon rate **c** is fixed at outset, with due regard for rates of return and economic conditions prevailing at the bond issue date. For example, in 2010 conventional British funds included those with coupon rates between 2¼% and 12%, issued at various past dates (see figure 7.3), and figure 5.15 shows interest rates and the annualised rate of change in output value flow **ΔG/G** in the UK and USA. At times both variables have been high, though not necessarily equal to each other. It is unlikely that investors will be attracted to purchase newly issued bonds that do not provide a return that at least equals, if not exceeds, the current rate of return of the day – unless the bond was issued at a discount.

It is logical therefore to assume that the coupon rate of a bond will likely be set to reflect a level of return in excess of interest rates and at or about expected output value growth. We are not saying that the coupon rate **c** always reflects the growth rate of output value flow. Clearly the latter varies a good deal over time. But we are saying that the rate set depends upon interest rates and economic conditions prevailing at the issue date, and thereafter fixed for the life of the bond. Making an assumption of this kind we could write in annualised, discrete terms for money:

$$i = \left(\frac{\Delta G}{G}\right) - \Delta s_{money}$$

and likewise for a bond with maturity of ξ years:

$$r_\xi = c - \Delta s_\xi \qquad (7.24)$$

Where r_ξ is the interest rate or yield to redemption and Δs_ξ represents the annualised bond entropy change or disequilibrium position with respect to the coupon rate of a particular bond.

Clearly entropy change in this equation is a little different to that in a short-term monetary situation, chiefly because of changes in economic conditions χ (in particular inflation) at the outset of a bond and during its lifetime. Figure 7.3, for example, shows how past coupon rates for British funds compare to their yield to redemption at a particular time.

It is, therefore, a matter of the extent to which the coupon rate set at the outset of a bond's life subsequently equates to interest rates and output value growth in the future, and as an alternative we could write:

$$r_\xi = (c - \chi) - \Delta s_{money} \qquad \text{or} \qquad (7.25)$$
$$r_\xi = c - \Delta s_\xi, \qquad \text{where} \qquad \Delta s_\xi = (\Delta s_{money} + \chi)$$

Even so, it is a fact that widely varying coupon rates *are* effectively 'normalised' to a yield curve as shown in figure 7.3, the changes being effected by individual entropy changes Δs_ξ reflected in the price N_P of each bond, with the price differing in a complex manner across the period ξ. Thus we can substitute equation (7.24) into the bond discount equation (7.19) and write for standard or exponential notation:

$$\frac{N_P}{\Phi} = \left[e^{-(c-\Delta s_\xi)\xi} + \frac{c}{(c - \Delta s_\xi)}\left(1 - e^{-(c-\Delta s_\xi)\xi}\right) \right] \qquad \text{or}$$

$$\frac{N_P}{\Phi} = \left[\frac{c - \Delta s_\xi e^{-(c-\Delta s_\xi)\xi}}{c - \Delta s_\xi} \right] \qquad (7.26)$$

From equations (7.24) and (7.26), we can see that when entropy change Δs_ξ is zero, bond price N_P is equal to the par value Φ, and yield to redemption r is equal to the coupon rate c.

When entropy change Δs_ξ is positive, then bond price N_P is greater than the par value Φ, and likewise yield to redemption r_ξ is less than the coupon rate c. Similarly, when entropy change Δs_ξ is negative then $N_P < \Phi$, and yield to redemption $r_\xi > c$.

Figure 7.5 sets out the postulated effects of entropy change on yield.

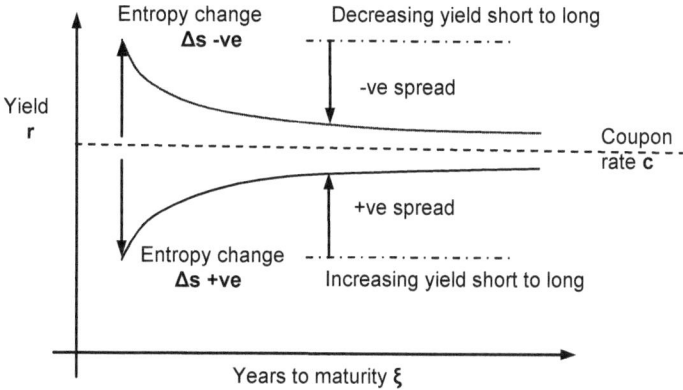

Figure 7.5 Postulated effect of entropy change Δs on yield to redemption r, assuming issued at par value

All of the above leads to the premise that entropy change likely plays a part in determining yield to redemption, at either above or below the coupon rate, and that the relationship depends upon the time period to redemption, and upon the level of the coupon rate compared to current interest rates.

The particular problem that we now have to deal with, which cannot be entirely explained in thermodynamic terms, is that of the time dimension of entropy change; how to visualise the shape and spread of an entropy function over the period to maturity of a bond, given that little or no knowledge of the future, say fifty years ahead, will be available. All that buyers and sellers have available to them is knowledge of cumulative events to date, and long term trends (including the shape of the yield curve at any point in time) combined with a picture of the short term.

This problem was touched upon when considering discounted cash analysis earlier in this chapter, whereby an exponential discount function was used to put a weight on future outcomes. Noting that our bond equation (7.26) contains negative exponentials, it might be expected that an entropy discount function would be similarly related, as was assumed for project and annuity investments. Sometimes, however, yield curves can have complex shapes, including humps and other non-linearities, so a decreasing exponential might not offer a 'fit all' solution. Frederick, Loewenstein and

O'Donoghue (2002) for example believe that the Discounted Utility model introduced by Samuelson (1937) has little empirical support and that consideration should be given to other models, including hyperbolic discounting. The empirical research of Nelson & Siegel, Anderson & Sleath and others already referred to utilises complex exponential functions combined with knowledge of short, medium and long-run interest rates to derive the instantaneous forward rate curve, as illustrated earlier at equation (7.22).

For the purposes of this analysis we will continue with the simple exponential model that was assumed earlier, to see where this leads us, though a more complex approach might be considered for future work. Figure 7.6 shows the yield curve from connecting the yields of conventional British Funds at October 2010 that were illustrated at figure 7.3 and along with it a declining entropy function, such that the addition of the two equals a constant level of an average current coupon rate.

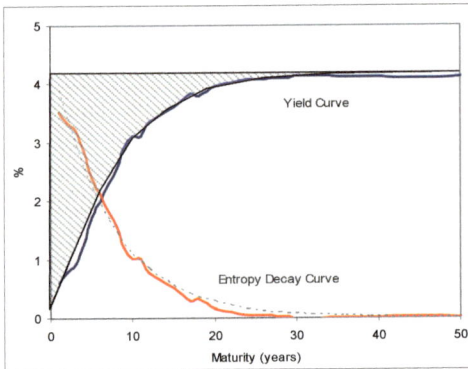

Figure 7.6 Yield curve and entropy function British conventional funds October 2010.

We might posit that the entropy function in this case could have the same form that we assumed earlier at equation (7.10):

$$\Delta s_\xi = \Delta s_0 e^{-\varepsilon \xi}$$

Where Δs_0 is the current entropy change level and ε is some form of decay factor, dependent upon investors view on the future. However, readers should not assume that such a function is axiomatic. A yield curve with a hump for instance will require a more complex function.

Nevertheless, pursuing the matter to a conclusion, the advantage of an entropy function of this kind, or similar, is that it can be *positive* or *negative* about the mean of the current coupon rate, resulting in a 'flip' of the curve, enabling the shape of the yield curve to go from 'upward-sloping' to 'inverted' and vice versa. It can meet situations where short-term economic trends change.

Similar to the approach to discounted cash flow analysis at equations (7.8) – (7.11), we could set out two equations for the yield, one for the short term, denoted by **0**, and the other for the long term, denoted by ξ, and then combine them as follows:

$$r_0 = c - \Delta s_0$$
$$r_\xi = c - \Delta s_\xi \qquad \text{giving:}$$

$$r_\xi = r_0 + \left(\Delta s_0 - \Delta s_\xi \right) \qquad (7.26)$$

The long-term yield to redemption of a bond then becomes a function of the short term rate of interest and the difference between the entropy changes associated with each. Thence by substituting in our entropy decay equation we can write:

$$r_\xi = r_0 + \Delta s_0 \left(1 - e^{-\varepsilon \xi} \right) \qquad (7.27)$$

which expresses the yield to redemption in terms of the short term interest rate, the short term entropy change and a decay function. And in the alternative, by turning this around we could also write:

$$\left(r_\xi - r_0 \right) = \Delta s_0 \left(1 - e^{-\varepsilon \xi} \right) \qquad (7.28)$$

Which expresses the yield spread between long and short-term perspectives, in terms of the short term entropy change and the decay function. It is emphasised again that further research on the shape of the decay function is required in respect of more complex yield curves.

It is of interest to compare the yield spreads between long and short dated bonds, set against money entropy change for the UK and USA economies (based on M4 and M2 money data respectively). The yield spread data obtained for a long time series was for 20-year BGS for the UK and 10-year bonds for the USA, both set against 3-month treasury rates. If one assumed a decay rate of 5% p.a., then the discount decay factors $(1 - e^{-\varepsilon\xi})$ applicable would be 1 for the UK and 0.5 for the USA. Figure 7.7 shows the results of this set of assumptions.

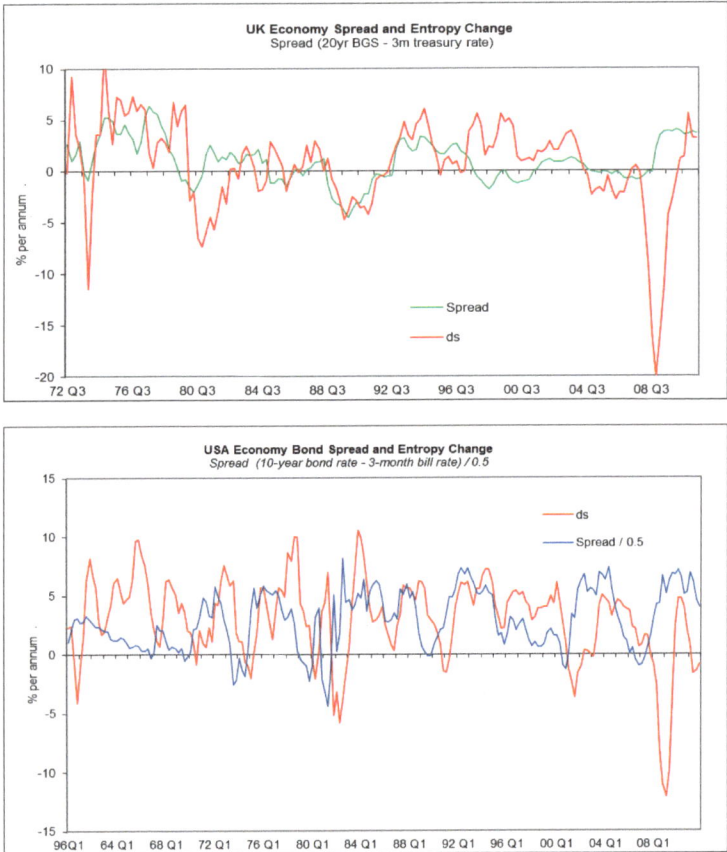

Figure 7.7 Modified Yield spread (quarterly) versus current money entropy change (4-quarter moving averages) for the UK and USA economies.

Statistical regression of the result does not produce a good correlation, though it can be seen that yield spreads appear to follow to some degree the ebb and flow of money entropy change associated with money. In particular, in the USA values of spread appear on occasion to act as a lead factor to economic entropy change. Further work is therefore required to examine this effect in more depth, in particular lags and leads of possible impacting economic factors, improved long-run modelling of the value capacity coefficient (see figure 5.18 and following comments) and more complex decay functions, though that is beyond the scope of the research covered in this book.

CHAPTER 8 ENERGY, RESOURCES AND THE ECONOMY

8.1 Energy and the Economy

Economic systems of developed and developing countries have become progressively embedded in an energy base, to provide a source of productive power and human wealth and well-being. Energy consumption (technically exergy consumption) provides electricity, powers machines in industry and computers, provides heat for industry and homes, and powers road, rail and sea transport. Fishing and agriculture (including the manufacture of fertilisers) in developed economies are now significantly dependent on energy, rather than human or animal power. Ayres and Warr (2004) have shown through the use of a LINEX function that exergy consumption can explain most of economic growth in the USA since 1900, with the marginal productivities (elasticity) of capital stock and labour (especially the latter), being well below that of the work output of exergy consumption.

Energy is now an international commodity, and few countries with a significant manufacturing and commercial base can now be described as 'closed' with respect to energy. USA oil production peaked in about 1970 and the USA now imports $2/3^{rd}$ of its oil requirements. Of the major economies, only Mexico, Canada and Norway can claim to be net exporters of oil. In the natural gas market, the USA and Europe are now net importers of gas via pipelines respectively from Canada and Russia. Only in the coal industry is consumption met mostly by local production, with China and USA accounting for 60% of world production and consumption. In 2006, ten developed countries with just 13% of world population (USA, Japan, Germany, Canada, UK, South Korea, Italy, France, Australia and Spain) accounted for 45% of world GDP, and 40% of primary energy consumption; with China, Russia and India bringing the latter total up to 66%.

By common practice, the units used to measure energy production and consumption are those of weight (tonnes of oil), volume (cubic metres of gas) or the heat value of sources of energy (Joules, BTU's). These can be equated to 'productive content' or exergy if account is taken of the net energy delivered to the environmental average.

Figures 8.1 and 8.2 summarise the development and relationship of primary energy consumption and electricity generation to GDP and population over several decades for some key countries.

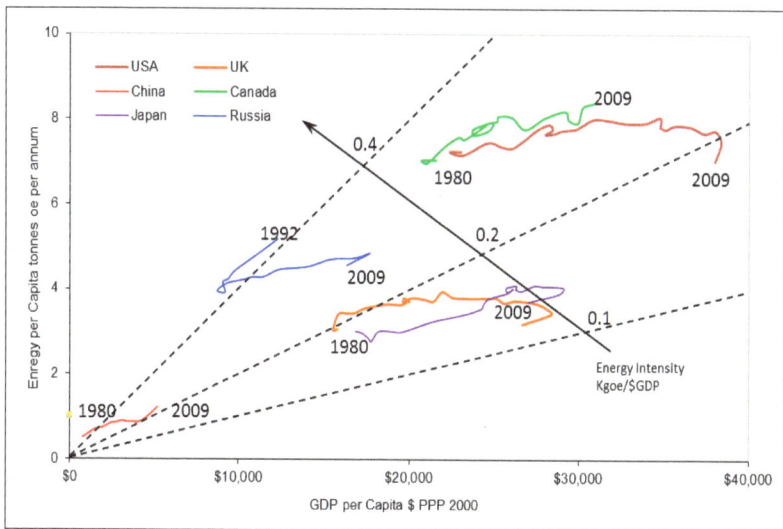

*Figure 8.1 Primary Energy Consumption per Capita versus GDP per Capita 1980 – 2009
(PPP 2000 levels) Russia 1992 onwards   Source: OECD, Penn World, BP*

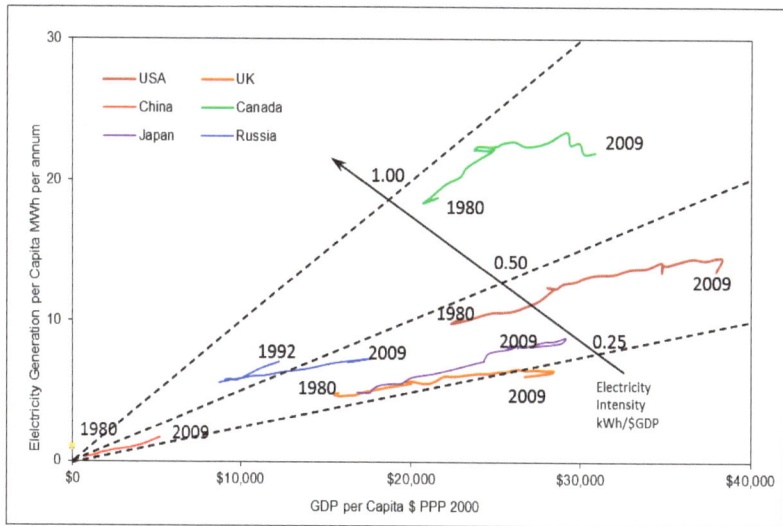

*Figure 8.2 Electricity Generation per Capita versus GDP per Capita 1980 – 2009
(PPP 2000 levels) Russia 1992 onwards   Source: OECD, Penn World, EIA*

In some developed economies, energy consumption per capita has begun to level off; indeed in the 2008 recession it declined significantly. But in the developing world, illustrated by China, energy consumption per capita continues to grow. Energy intensity (energy/GDP) has fallen significantly for all countries, and continues to do so, but has funnelled down to a narrower range than three decades previously. However, bearing in mind that electricity production accounts for some 40% of primary energy consumption, it should be noted that per capita electricity consumption is still rising, associated with only a small decline in electricity intensity (ratio electricity generation/GDP). Thus a significant part of the reduction in energy intensity has been owing to improved efficiency in conversion of energy into electricity, though it should be cautioned that ultimately the laws of thermodynamics place limits on the level of efficiency that can be obtained. The International Energy Agency indicates [IEA Worldwide Trends in Energy Use & Efficiency (2008[1])] that potential for further savings in primary energy consumption could be between 18%-26% in industry, and 23%-32% in electricity generation.

8.2   Energy and Capital Stock

The use of energy in an economy is predominantly through the powering and installation of associated capital stock, such as transport, machinery and computers, and the provision of heat, though some oil finds its way into products such as chemicals and fertilisers. Table 8.1 for example sets out an analysis of US capital stock and consumer durables.

A significant proportion of assets is directly associated with energy consumption, such as equipment & software and structures in power, communications, steel, petroleum, mining, railways, road vehicles and consumer durables; approaching a quarter of all fixed assets. Without an energy supply these would cease to have a function. Even farming is now energy intensive in the Western world. A further point to note is that humankind now consumes energy in order to make most fixed assets. Modern buildings factories and roads are made by using construction equipment and cranes, and machinery and vehicles are made by consuming energy through manufacturing plant & machinery. A return to human and animal power, predominant before the 20th century, might be viewed by *Developed Economic Man* as being unimaginable and a retrograde step. Such a step would certainly be dramatic.

Table 8.1 US Current-cost Net Stock Fixed Assets & Consumer Durables
(www.bea.gov Fixed asset tables 3 and 11b)

$bn 2007	Private	Government	Total	%
Equipment & Software				
Computers/Software	$623.5	$46.4	$669.9	1.4
Communications	$555.4		$555.4	1.2
Medical Equipment	$269.2		$269.2	0.6
Office Equipment	$188.3		$188.3	0.4
Engines Turbines	$83.5		$83.5	0.2
Electrical transmission	$358.4		$358.4	0.8
Industrial M/c	$1,041.0		$1,041.0	2.2
Trucks/buses	$794.8	$24.1	$818.9	1.8
Autos	$141.4		$141.4	0.3
Aircraft, airborne equipment	$296.5	$160.0	$456.5	1.0
Ships, boats	$65.4	$142.3	$207.7	0.4
Railroad equipment	$101.5		$101.5	0.2
Agricultural M/c	$147.0		$147.0	0.3
Construction M/c	$149.8		$149.8	0.3
Mining/Oilfield Equipt	$49.5		$49.5	0.1
Other	$470.8	$524.2	$995.0	2.1
Sub-total	$5,336.0	$897.0	$6,233.0	13.4
Structures				
Residential	$17,819.1	$323.7	$18,142.8	38.9
Offices	$1,620.8	$627.1	$2,247.9	4.8
Commercial	$1,834.1	$34.0	$1,868.1	4.0
Hospitals	$706.0	$225.9	$931.9	2.0
Manufacturing	$1,175.9	$66.9	$1,242.8	2.7
Power	$1,230.6	$241.4	$1,472.0	3.2
Communication	$485.8		$485.8	1.0
Petroleum/Nat Gas	$832.6		$832.6	1.8
Mining	$57.2		$57.2	0.1
Railroads	$298.7		$298.7	0.6
Farms	$307.8		$307.8	0.7
Highways/Streets		$2,634.1	$2,634.1	5.7
Military		$391.4	$391.4	0.8
Transportation		$532.4	$532.4	1.1
Educational	$362.6	$1,608.6	$1,971.2	4.2
Sewers & Water		$912.0	$912.0	2.0
Other	$1,313.9	$716.4	$2,030.3	4.4
Sub-total	$28,045.1	$8,313.9	$36,359.0	78.0
Consumer Durables				
Autos	$573.1		$573.1	1.2
Trucks	$759.0		$759.0	1.6
Videos/Computers TV	$793.5		$793.5	1.7
Appliances	$204.2		$204.2	0.4
Other	$1,695.7		$1,695.7	3.6
Sub-total	$4,025.5		$4,025.5	8.6
Total Assets	$37,406.6	$9,210.9	$46,617.5	100.0

8.3 Economic Output

To assess the impact of changes in energy availability on economic output, we return first to equations (2.4) and (2.9) at chapter 2 where, *at a specific moment in time,* output value G_M in current money terms (the GDP) was deemed to be *equivalent,* but not equal, to output value G_O in terms of productive content (the real value inside output) net of efficiency losses.

To compare output over time with volumetric aspects of product inputs and outputs, economists calculate a price *index* in terms of a weighted average of prices of relevant goods, albeit that variations in exchange rates can further complicate the issue. The output 'volume' obtained from using a price index therefore hopefully nets out changes in value imparted to it by economic pricing. Where possible of course, economics makes use of real volume, weight and energy value measures of commodities, energy and foodstuffs to add substance to the calculation. Such calculations become less easy to measure and more divorced from a base, however, the further up the economic chain one moves towards output values attributed to humans that provide a service on the back of output, effectively overheads.

Given that output value in terms of productive content is made up of the *true* contributions of each of capital stock, labour and inputted resources, and not their money equivalent, as per figure 2.5, chapter 2, then variations in output value in terms of real productive content might be more affected by changes in non-human factors, than might otherwise be assumed.

Further, as already shown in this chapter, since energy resource availability currently constitutes perhaps the major input to economic output, then the workings of the international energy market will have an important bearing on economic output. At a local level, one has only to think of a strike by oil delivery truck drivers to witness the 'panicking' effect on consumers looking for assurance for the immediate future. At an international level of course the effect is far more marked; for example, the large rises in oil prices in 1974, 1979, 2005-2008 and 2010-2011, as shown at figure 8.3, where each rise was subsequently followed by a world recessionary/ deflationary process, reducing demand.

The thermodynamic explanation for this effect is that active output volume change dV/V came up against a greater resource constraint change dV_c/V_c, which enacted a significant negative economic entropy change ds (a dis-utility and exit of entropic value dQ).

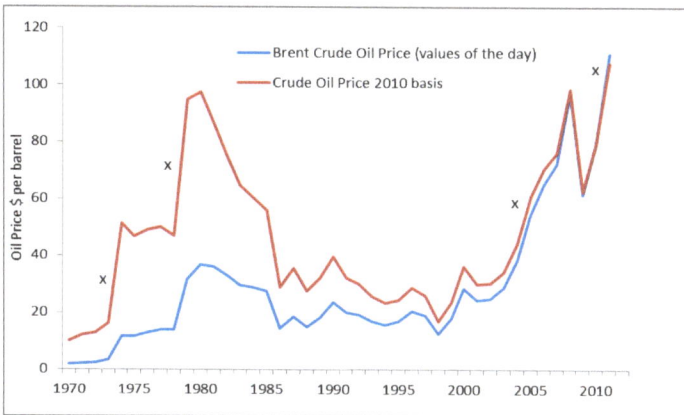

Figure 8.3 Crude Oil prices – Source BP Statistical Review 2010, IndexMundi, Bloomberg, OECD.

The impact of this is to occasion first a rise in oil prices, as in figure 8.3, and then a drop, as demand turns down to meet the available output. It cannot be imagined that the wage earners of exporting entities engaged in the production of oil had suddenly acquired a much larger real impact via their input of productive content; more likely that productive capacity at that time was constrained, compared to demand, creating an economic entropy loss and a shortage of energy product to the international market.

The net effect of a change to output value of a factor such as energy is reflected in its interaction with all the other factors that make up GDP. We are not saying that output values of non-human energy factors alone determine economic activity; clearly human and other factors matter too, but we are saying that energy consumption now constitutes a significant input to the economic process as it is currently constructed.

8.4   Non-Renewable Energy

According to the US Energy Information Administration, oil, natural gas and coal accounted for 37%, 23% and 27% respectively of total world primary energy consumption in 2005, with the remaining 13% coming from nuclear, hydro, wind and other sources. Given the long timescales associated with nuclear technology, it is likely that consumption of oil, natural gas and coal will remain important factors in determining levels of

GDP in the short to medium term horizon, though with some decline in energy intensity.

Figure 8.4 is a diagrammatic representation of a non-renewable resource, such as oil, gas and coal, similar in construction to the Logistic curves set out at figure 4.7 in chapter 4, and described by equations (4.30) and (4.31).

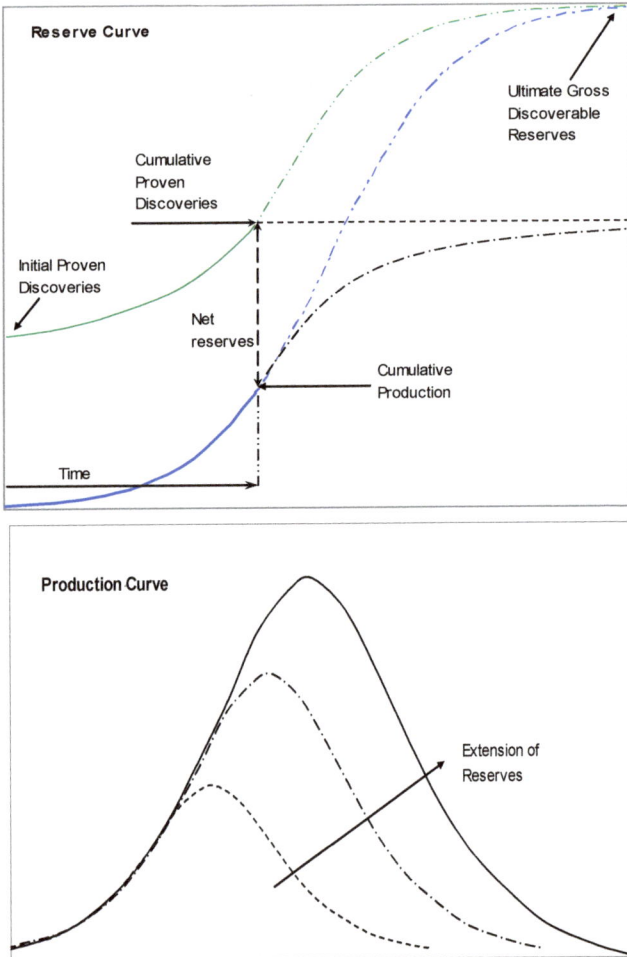

Figure 8.4 A Non-renewable Energy Resource Reserve

At a current point in time, information is known of production (depletion) rates and cumulative production to date. Knowledge of remaining net proven reserves that are being tapped will also be known from geological data. What is not known for certain is what other reserves may be discovered or become recoverable in the future. Oil and gas professionals have techniques at their disposal to estimate future discoveries. One such technique is creaming, which is based on the knowledge that the easiest energy fields are those that are discovered first and the more difficult ones later [International Energy Agency (2004)]. Other experts extrapolate from current production using the Hubbert equation [M King Hubbert (1956)], which is based on the Logistic/Verhulst equation set out in chapter 4.

In respect of oil, there is a significant element of political positioning surrounding some countries' estimates of what are declared as proven reserves, in particular those of some members of OPEC, such as to place some doubt as to the reliability of such reserve estimates. Laherrere (2005, 2006) suggests 'backdated mean technical reserves' as a measure independent of oil companies and national agencies' official figures. He concludes that since 1980 oil discoveries have been less than production growth. Groups such as the Energy Watch Group and the Association for the Study of Peak Oil & Gas also have doubts about the official position.

Figure 8.5 summarises oil and gas discovery and production levels per annum as a per cent of net reserves. Calculations of the per annum rates of discovery are based on the annual changes in the official figures for net proven reserves [taken from the BP Statistical Review], and then adjusted by annual production levels. Should reserves subsequently be reassessed either upwards or downwards, in particular those for oil, then the figures for discoveries would incorporate an appropriate adjustment upwards or downwards, causing the curves at figure 8.5 either to draw apart or to cross.

If cumulative production proceeds against a finite reserve, there comes a turning point when production peaks. This is at about the middle of the graph at figure 8.4 - the so-called 'peak oil' phenomenon. Thereafter production begins to wind down as the remaining reserves reduce. Such a peak was reached in the USA oil industry in about 1970. The only way in which production can increase in perpetuity is by the discovery of new energy resource fields - of ever increasing size and numbers.

However, knowledge of the ultimate level that proven reserves may reach is not known for certainty, and therefore estimating the peak production level

186

and when it may occur is an exercise in trend analysis, sensitive to determining the turning point.

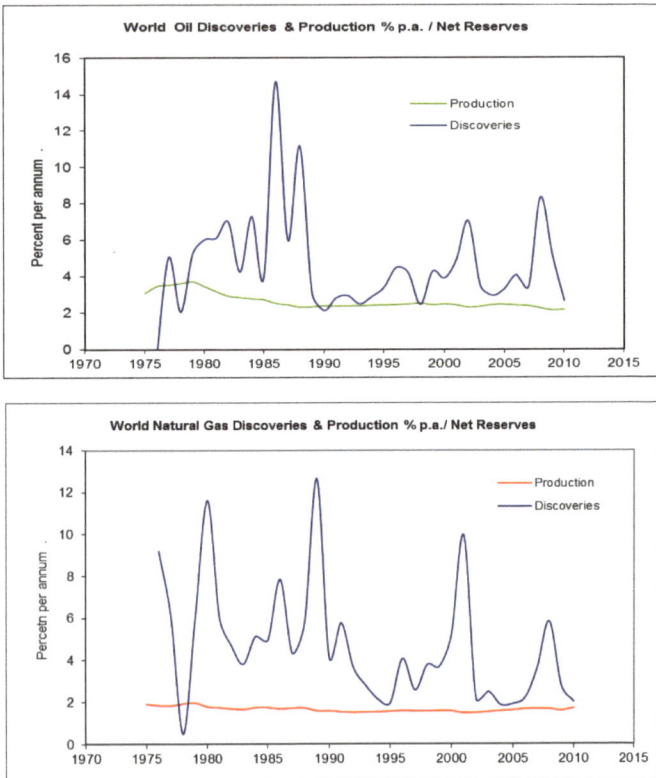

Figure 8.5 World Oil and Natural Gas Discoveries & Production % p.a. / Net Reserves
Data source: BP Statistical Review 2000 – 2011

There is much debate about when world 'peak oil', 'peak gas' and 'peak coal' may occur, depending upon views held on the ultimate level of proven usable reserves, where the level of exergy to be abstracted is greater than that put in to develop such reserves.

Figures 8.6 and 8.7 set out charts of world cumulative production and official net proven reserves for both oil and natural gas. Cumulative world oil production as a proportion of official net proven reserves remaining (not including tar sands) continues to rise, and is approaching a 1:1 ratio, indicating perhaps that production is nearing the peak illustrated at figure 8.4.

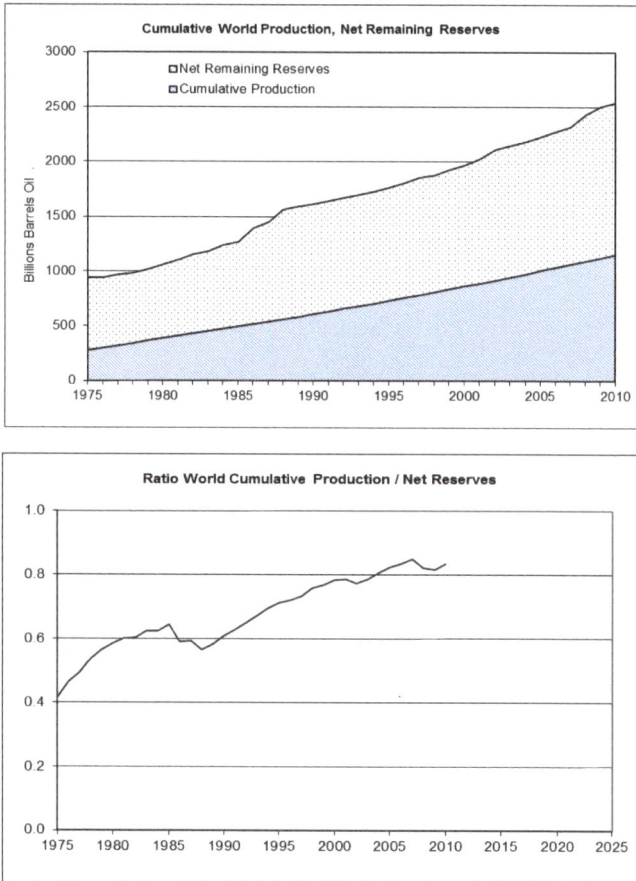

Figure 8.6 World Oil Reserves and Cumulative Production   Sources: BP, EIA.

On present trends, based on oil companies' and national agencies' estimates of net reserves, peak oil may occur in about a decade. On more pessimistic estimates of the level of net reserves, the peak may already have arrived, with production continuing flat for a while before reducing. Estimates of ultimate recoverable reserves range from 3.5 trillion barrels of oil [IEA WEO (2008)], down to 3 trillion [Laherrere (2006)], 2 trillion excluding extra-heavy oil. The position with natural gas appears to be further behind oil, though it is a matter of evidence as to whether gas reserves yet to be discovered will occur.

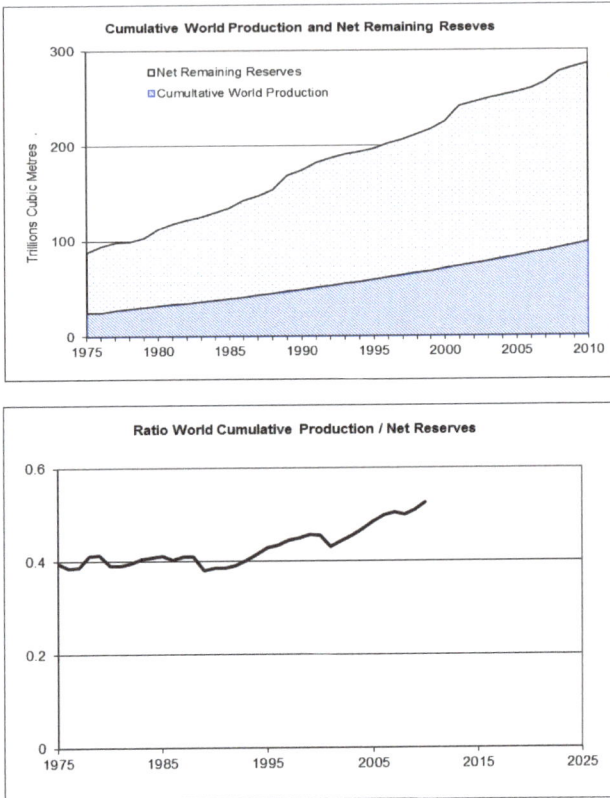

Figure 8.7 *World Natural Gas Reserves and Cumulative Production*
Sources: BP, Laherrere ASPO

A more recent possible entrée to the gas market is the technology of hydraulic fracturing or 'fracking'; injecting a fluid made up of water, sand and various chemicals (some toxic) deep into the ground, causing nearby shale rock to crack, creating fissures for natural gas to collect. Such a technology could undoubtedly unlock substantial additional gas reserves, though opponents to the technology cite significant environmental, safety and health hazards, such as water and air contamination, and potential earthquakes.

Figure 8.8 summarises world oil and natural gas production and consumption. The shape of the production charts corresponds to a part of the left-hand side of the production curve at figure 8.4. It is a matter of debate as to when production, and thereby consumption of oil and gas, may peak and begin to decline.

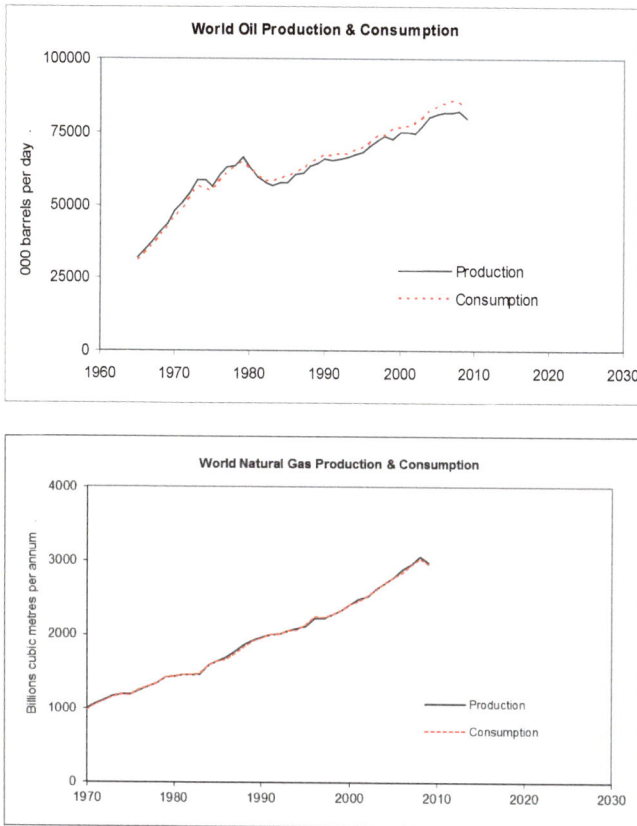

Figure 8.8 World Oil and Natural Gas Production and Consumption Source: BP

Of course, if new proven reserves of oil and/or natural gas are found, of sufficient economically extractable volume to increase significantly the ratio of net reserves to cumulative production (depletion), or real improvements are obtained for recovery factors applied to 'oil in place', or a severe recession occurs to reduce demand for energy, then peak production may be put off further, or production may flatten off at a higher level for a period. On the other hand, if some reserve estimates are subsequently brought down, in particular those of OPEC referred to earlier, then the depletion curve would be brought forward. On the assumption, however, that a peak will eventually be reached, it might be expected that production might then gradually decline in level over many years, corresponding approximately to the right-hand side of the production chart at figure 8.5.

Besides oil, and natural gas reserves, there is also some debate as to the extent of world coal reserves. According to the BP Statistical Review 2011, total world proven coal reserves stood at 860 billion tonnes of coal; split about half bituminous/anthracite and half lignite/sub-bituminous - the latter being of inferior quality to the former. This appears to represent a reserves-to-production (R/P) ratio of 118 years, based on 2010 production levels. Approximately 4/5th of these reserves are held by six countries: USA, Russia, China, Australia, India and South Africa. Coal is not widely-transported worldwide, being predominantly consumed in the country where it is produced. China and USA account for 60% of production and consumption. A report of the Energy Watch Group *[EWG-Series No1/2007]* concludes, however, that data on coal reserves is poor, the data for China for example being last estimated in 1992, even though 20% of reserves in that year have since been consumed. The report points out that worldwide coal reserves have been downgraded significantly, and estimates that global production may continue to increase over the next 10-15 years, reach a plateau, and then gradually decline. Such a picture is not dissimilar to the curves for oil and gas, though perhaps set later.

Figure 8.9 illustrates the reserves to production ratio since 2000. It has dropped by 45% in only a 10–year period, and looks set to fall further, if reserves continue to fall and production to rise. By comparison, the current reserves/production ratios for world oil and world natural gas stand at 46 years and 59 years respectively.

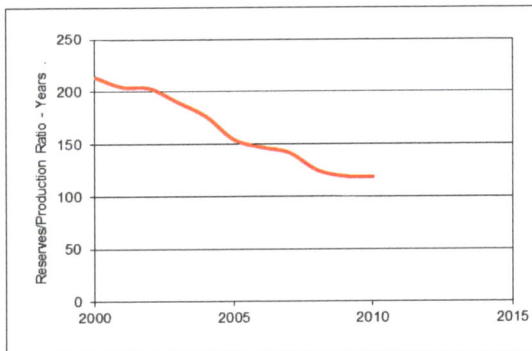

Figure 8.9 World Coal Reserves/Production Ratio – Years
Data source: BP Statistical Review 2000 – 2011

It is unlikely that a reduction in current energy resource production will wholly feed through into a reduction in GDP. There are other effects to consider, such as substitution of other energy sources, new reserves and further improvements in efficiency of use affecting energy intensity. But it is the case that relative energy dependency and the path of primary energy input are likely to play key roles to any future position concerning economic development.

8.5   Energy Resource Substitution

Climate change apart, making the assumption that in due course some non-renewable resources are eventually rendered inert through irrevocable consumption, the principle of maximisation of entropy gain, set out in chapter 4, is then invoked. Humankind, having gone along one branch of the stream to discover a potential dead-end, will gradually seek other avenues to advance its cause, by turning to alternative sources of energy.

Nevertheless, there may still be some gains to be made from improving efficiency, thereby increasing the net exergy of conventional non-renewable energy sources delivered to economic output (subject to limits set by the Laws of Thermodynamics), and compared to the reductions in energy intensity achieved over the last 40 years (see figure 8.1); for example, by ground transport increasingly switching to electric energy, entailing changes to the capital stock.

Further large improvements and maintenance of positive economic entropy gain will likely only be achieved by finding substitutes for conventional energy resources, such as nuclear power, and renewable sources such as solar, wind and hydro generated electricity. Potentially, however, these may involve higher economic costs, and therefore lower potential net volume benefit to economic output. Wind power will also require stand-by power plant when wind force is low.

Nuclear power, although having low operating costs, entails high capital costs, and high waste and decommissioning costs, with waste potentially accumulating much beyond many generations of humankind. Nuclear capacity carries with it also some risks of safety, though these can be controlled. Figure 8.10 summarises world electricity consumption from nuclear power up to 2010. After the rise of the 70s and 80s, output growth began to slow down and has been largely flat since about 2004.

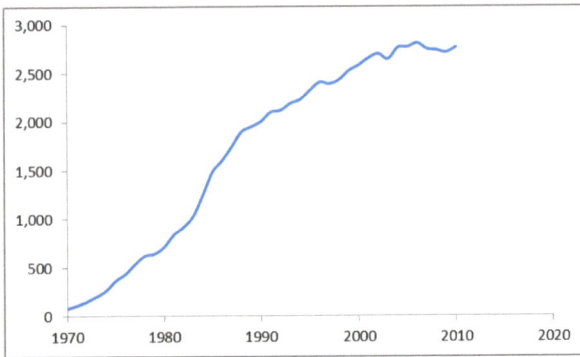

Figure 8.10 World Nuclear Electricity Consumption TWh   *Source: BP*

Major countries that now have a high percentage of installed nuclear power capacity include France (78%), Belgium (55%), Sweden (52%) and Ukraine (51%) with a number of other European countries above 30%. The top five capacities in 2001 were USA (102,000 MWe), France (62,000 MWe), Japan (46,000 MWe), Russia (22,000 MWe) and Germany (22,000 MWe) [Source: EIA/IAEA].

Figure 8.11 confirms that the nuclear industry has lost share to other forms of electricity generation since the early 1990s.

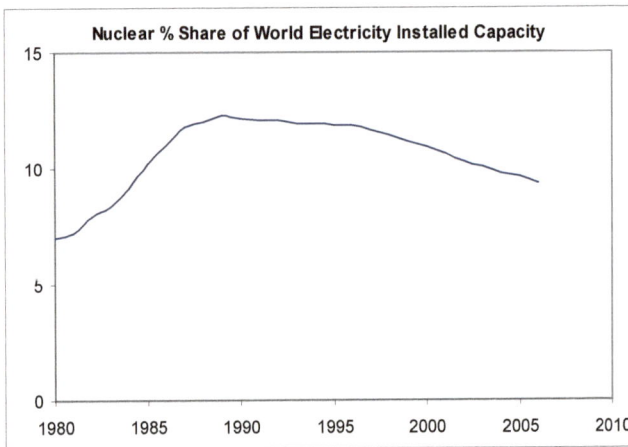

Figure 8.11 World Nuclear Power Installed Capacity   *Source: EIA*

Growth of hydroelectric power (a renewable resource) is limited by geological/landscape considerations, and its share of electrical output has marginally fallen to about 16% [source: BP]. By contrast, figure 8.12 indicates that the share of electrical generation capacity taken by other renewable energy sources, such as geothermal, solar and wind power, is growing, though it still commands only a small share.

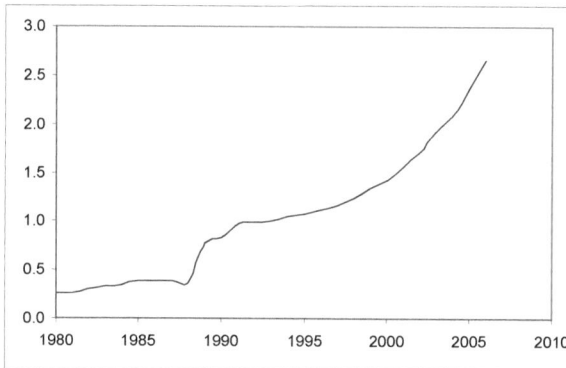

Figure 8.12 % Share of World Electricity Installed Capacity taken by
Geothermal, Solar, Wind, and Wood and Waste,   *Source: EIA*

Accepting that an ultimate limit on fossil resources may exist, then the alternatives to energy substitution are either a quantum shift in consumers' attitudes towards energy use, removing a part from GDP and permanently altering the mix of output, or a long-term levelling or decline in GDP, which may lead to a reduction in GDP per capita, unless population also declined in proportion.

8.6   Thermo-economic Considerations

Although new reserve discoveries for a resource make no difference to the eco-system as, by definition, they were there in the first place, they do make a difference to an economic system by virtue of the potential economic gain to be maximised. Potential economic entropy gain occurs as more and more resource is discovered (that is a utility value is created). Once discovered and developed, such potential gain is then gradually reduced as the reserve is depleted through production. During production and subsequent consumption, real thermodynamic exergy is consumed and thermodynamic entropy increases arise.

Dealing with the path of economic entropy potential to develop and *deplete* a limited reserve, from equations (4.30) and (4.31) at chapter 4, the standard model for a non-renewable resource is that of the Verhulst equation, and output volume **V** from a reserve could be stated as being equal to a function of the change in accumulated production **N** per unit of time:

$$V = f\left[\frac{dN}{dt}\right] = \varphi R b (1 - b)$$ (8.1)

With the solution at any time **t** of:

$$N(t) = \frac{N_0}{\left[\left(N_0 / R\right)\left(1 - e^{-\varphi t}\right)\right] + e^{-\varphi t}}$$ (8.2)

Where φ is the production/net reserves rate per annum, **R** is gross proven reserves, $b = N/R$ is the ratio of cumulative production **N** to proven reserves **R**, and N_0 is a starting position. By differentiation it can be proved that maximum output value occurs at the point $b = N/R = \frac{1}{2}$; that is at the mid-point of the production and reserve curves at figure 8.5.

As with the chapters on money and the labour market, we imagine a constraint to output, which is a function of the total resource depletion **N** that has been used up. The higher this is the less incentive there will be to exploit a resource reserve. We could express the potential output volume flow no longer available as:

$$B = \varphi R b$$ (8.3)

By comparing equations (8.1) and (8.3) it can be seen that:

$$V = (1 - b)B \quad \text{or} \quad \frac{V}{B} = (1 - b)$$ (8.4)

By taking logs and differentiating equation (8.4) we have:

$$\frac{dV}{V} - \frac{dB}{B} = \frac{-db}{(1 - b)}$$ (8.5)

From our volume economic entropy equation (4.23) at chapter 4 we could write:

$$ds_{depletion} = k\left[\frac{dV}{V} - \frac{dB}{B}\right] = \frac{-kdb}{(1-b)} \qquad (8.6)$$

where $ds_{depletion}$ is the potential *economic* entropy change accompanying the volume depletion of units reserve, and **k** is the productive content of each unit (replacing money in this equation).

Figure 8.13 illustrates the loci of all the resource depletion factors, assuming **k=1**.

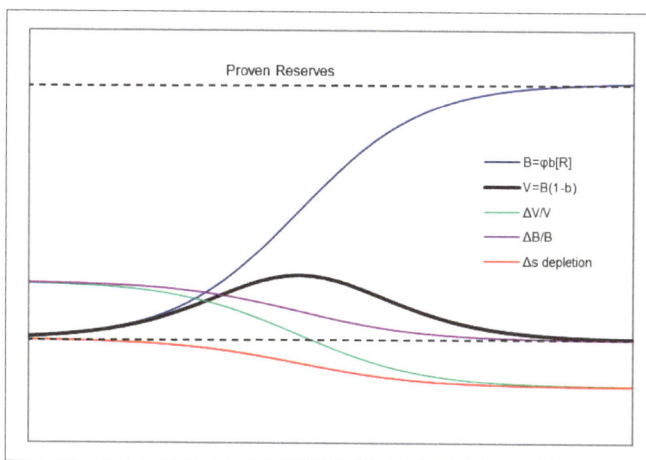

Figure 8.13 Loci of cumulative resource depletion: Resource constraint B, output volume flow V, rate of growth of output value flow ΔV/V, rate of growth of resource constraint ΔB/B and unit economic depletion entropy change Δs.

Resource constraint **B** follows cumulative resource depletion **N (where N = bR),** being an S-shaped curve, as per equation (4.31) for the Verhulst equation, and as illustrated at figure 8.5. The constraint eventually reaches proven reserves **R** when all the reserves have been abstracted. Output volume flow **V** from the reserve to the economic system follows a bell-shaped curve, reaching a maximum at the halfway point. The rate of change of output volume flow **ΔV/V** starts off at a maximum, based on the resource available to be extracted, but gradually declines to zero at the halfway point, and thereafter the rate of change becomes negative as output volume flow **V**

196

reduces. The rate of change of resource constraint $\Delta B/B$, starts off at a maximum, but gradually declines to zero once the resource has been used up. The net potential economic entropy change that arises from depletion $\Delta s_{depletion}$ [equal to $k(\Delta V/V - \Delta B/B)$], starts off at zero but gradually declines as the effect of the resource depletion takes hold, eventually resulting in a decline in the output growth rate $\Delta V/V$. It will be noted that, from the beginning of the development of a reserve, the rate of growth of the constraint is always greater than the rate of growth of output, though the effect is very small initially, not affecting economic entropy change or output growth significantly.

Should additional reserves be discovered or reserve recovery factors be improved, these will create a new economic entropy gain, extending the lifetime of the resources and arresting the impact of the decline in economic entropy change arising as the resource constraint builds up. Thus the production rate V is advanced, and the depletion rate B is set back.

We can therefore summarise net resource potential *economic* entropy drive as being equal to that arising from discovery less that which is subsequently lost to output via depletion:

$$\Delta S_{economic} = \Delta S_{discovery} + \Delta S_{depletion} \tag{8.7}$$

Where $\Delta S_{depletion}$ is negative. Once depleted, further economic entropy drive to be gained from the resource is zero, unless new discoveries are made.

In respect of output volume flow V, as exergy value is transferred from the resource reserve to the resource product in the production process, a Second Law loss occurs as costs of mining and transport are incurred, though some of this is picked up as other products bought by wages and profits. There is however an irrevocable loss of thermodynamic exergy and realisation of main body entropy, since, once the resource is mined, the exergy cannot be returned from whence it came – the ground or undersea – without expenditure in excess of that incurred in mining.

Last, as the product is consumed by the consumer, much of the exergy passed to the product in the production process is consumed and realised as an entropy increase to the environment. A total resource main-body entropy change can therefore be postulated of the form:

$$\Delta S_{Total} = \Delta S_{production} + \Delta S_{consumption} \tag{8.8}$$

Not included in the above representation is the effect of variable costs in developing a resource. For instance, the cost of exploiting Athabascan tar sands in terms of net extracted exergy is considerably more than extracting oil from most traditional oil wells.

Turning now to renewable resources, which could include food sources (fish, vegetation and animals, and the land and sea from whence they came), as well as other energy sources, the major difference compared to a non-renewable resource is that an input is provided by nature, the eco-system and the sun to re-supply the resource as it is used up, and entropy increases from production and consumption are counterbalanced by those emanating (eventually) from the sun through the eco-system. This is similar to the concept of new resources being discovered, set out in the previous analysis.

The standard model from which these can be derived is the *Lotka-Volterra* model [Alfred Lotka 1925, Vito Volterra 1926], where humankind is effectively the predator (**y**), and one or more of the other factors are the prey (**x**). On a volume basis they are linked via the differential equations set out at equation (8.9):

$$\frac{dx}{dt} = \alpha x - \beta xy \qquad \text{and} \qquad \frac{dy}{dt} = \delta xy - \varepsilon y \qquad (8.9)$$

In the model the prey is assumed to have an unlimited food supply, and to reproduce exponentially at the rate of **α**, unless consumed by predators, represented by a function **β** of the contact **xy** between predator and prey. Likewise growth in predators is governed by a function **δ** of the contact **xy** between predator and prey, and the death rate **ε** of the predators.

We will not develop further the mathematics here, since this is involved. Essentially, however, predators are dependent upon their harvest of prey, and prey are dependent on the sun, nature and the eco-system to reproduce themselves. If the predators have a high consumption rate of prey (in the case of humankind utilising also capital stock, energy and technology) such that the prey is gradually reduced, the sources on which the predators depend upon reduces, and the predators face a decline in population. The most obvious examples of this effect are over-fishing, over-farming of arable land and deforestation. Using up resources (many environmentalists might say squandering in an unsustainable way) entails a potential degradation of the earth's resources in the short-term ecological timescale.

In respect of renewable energy sources such as wind and solar power, the effective stock x is very small, with a high throughput. Thus energy generated is almost immediately transmitted to become electricity consumption by humankind.

Neither wind nor solar power is guaranteed in intensity, however, the first because of the variable, on-off speed of wind, and the second because of potential cloud coverage and night periods restricting hours of availability. However, siting solar power collectors in areas such as the Sahara desert might in part get around the latter [Mackay (2008)] to provide significant amounts of more continuous electrical energy. Methods of storing power could include, at appropriate times, the use of 'smart' chargers connected to batteries of electric cars directed to download power, and pumped water storage in hilly areas to convert potential energy into electrical power.

8.7   Population

Associated with the benefits of economic growth and energy consumption, there has been an increase in world population. It is a matter of opinion as to whether population size may now be approaching or has passed a sustainable ceiling within a finite world, and whether system constraints are building up to limit population size or reduce it. While humankind is not subject to any other predators, it is dependent upon prey and harvests, and its survival does depend upon on how it interacts with the eco-system, in a manner which had previously not been regarded as important.

A thermodynamic viewpoint of population growth differs significantly from the traditional economic approach, by virtue of evaluation of the real contributions to output value.

Thus a traditional economist might project output as a combined wage and profit rate w per unit of population, multiplied by population N, giving output value wN – supposedly nothing to do with the eco-system. The inference of this approach is that as the population increases, so the net new members are able to generate wealth to increase the GDP at an exponential rate, with the implicit assumption that the population is the source of wealth generation, and that resources are just there to be employed in that generation.

The thermodynamic approach, however, assumes that output value arises from the true contributions of each component – mostly resources and natural capital, and not economic man (though the latter is the instigator). Therefore a potential rise in output value will only occur if the productive content of the resources is there to form output.

In a position of resource shortage, or alteration of the eco-system arising perhaps from human activity, humankind likely cannot burn its way out to feed its growing population by consuming yet more, it has to consume less, witness past recessions following oil price hikes. Humankind, in increasing parts of the world, is a voracious consumer of resources for purposes other than for propagation of the species. Thus cumulative population growth potentially can become a constraint feeding into the entropy maximisation principle, which may in turn affect population size.

Similar to resource models, models of population incorporating a carrying capacity constraint are commonly based on the Verhulst formula set out at equation (8.10):

$$\frac{dN}{dt} = rN\left(1 - \frac{N}{R}\right) \tag{8.10}$$

Where **N** is population, **r** is the Malthusian parameter (the rate of maximum population growth) and **R** is the so-called carrying capacity (the maximum sustainable population). This equation does not necessarily indicate that population will eventually maintain itself permanently at a fixed level. If a decline occurs in resources available to support a population, this will affect the maximum sustainable population level **R,** which may come down, bringing about a decline in population size. If resources commanded rise, population levels may be sustained or rise.

There are indications that human birth rates in some areas of the world are declining. Between 1980 and 2010, population growth in West Europe has been only 11%, and the rest of Europe 5%. However, growth elsewhere has been larger – North America 35%, Asia 44%, South America 63%, and Africa 112%. In China measures to control family size have been in place for some time. The Russian population is currently declining. The US Census Bureau projects a world population level of 9bn by 2042, 31% more than the current level of 6.9m.

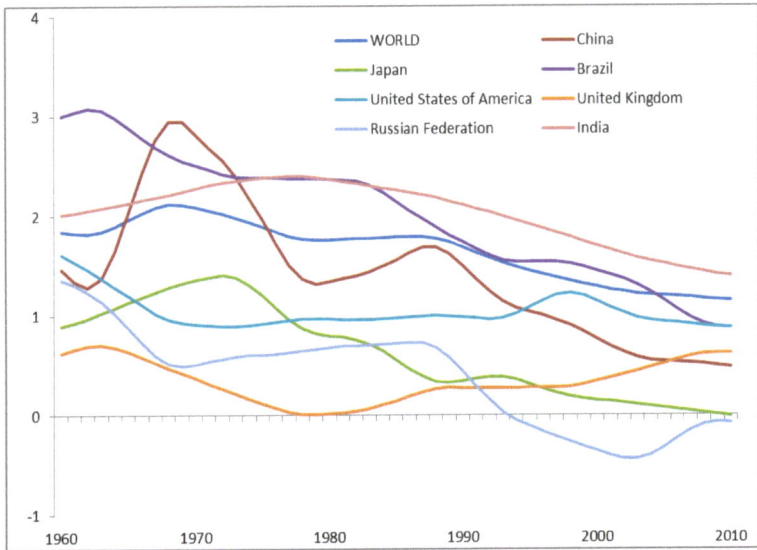

Figure 8.14 Population growth rates in selected countries and the world - percent per annum
Source: UN

8.8   Climate Change

Studies carried out by the Intergovernmental Panel on Climate Change *[IPCC]* and other agencies all point to the necessity of reducing or containing the impact of greenhouse gases on the world environmental climate and atmospheric temperature, though there continues to be some debate on the probable outcome. The scenarios painted by the agencies do, however, illustrate a point of this book, that restrictions placed on economic systems can potentially affect their forward path in quite a dramatic way, retarding the equilibrium position. The impact of the analyses presented to a world audience by IPCC may enact feedback to construct policies to change the shape of the future. In the case of climate change, the restriction arises on both sides of equation (4.3) increasing the costs associated with a waste function, and restricting output value based on fossil fuels.

There is of course nothing initially to prevent world economies continuing along a *'same as usual'* path, as they appear to have done so far, but without policies in place to reduce greenhouse gases, the probability is that air temperatures and sea levels will rise, to the potential detriment of the eco-system. Whereas, from the earlier resource analysis, any limits in oil, gas and coal reserves will ultimately automatically feed back into reduced fossil-fuel energy consumption, no such automatic feedback is likely with regard to achieving the required reduction in emissions. Instead, action to avert a rise in global air temperature is dependent upon agreement and co-operation among the international community. It is possible that such action might also be of an iterative, delayed kind, in response to accumulating environmental disasters, rather too late according to IPCC.

Actions by governments on behalf of the populace aimed at choosing a path which maximises potential entropy gain will therefore meet a constraint operating on two fronts.

First, if fossil-fuel burning is reduced, this will impact on the volume level of the GDP that is dependent upon this source of value generation – a large slice of the economic structure of the developed world. Follow-on to this policy is a change in direction of potential entropy maximisation, first towards other sources of energy not involving CO_2 production, and second by endeavouring to render fossil sources harmless (though the latter does not avert consumption of non-renewable sources). Both actions will involve significant additional, non-recoverable costs, which will add to the climate change constraint, resulting in a loss of economic entropic driving force. We could write

$$ds = k\left[\frac{dV}{V} - \frac{dZ_{CC}}{Z_{CC}}\right]$$

(8.11)

Where $\mathbf{k}$ is a composite productive content of output, and $\mathbf{Z}_{CC}$ is the constraint acting on that output. Clearly a large change in constraint upwards would reduce the economic entropy driving force $\mathbf{ds}$ significantly, impacting on output and causing it to slow down or decline.

If countries choose not to reduce fossil-fuel burning, on the evidence of IPCC, a significant rise in global air temperature and water levels will occur, which will cause a reduction in output value to some if not many economies through a loss of agricultural farming capacity, and a loss of habitation located in low-lying areas liable to sea flooding, with a consequent large-scale write-off of capital stock. Of more concern, however, is the potential significant effect on the eco-system, which humankind has previously mostly ignored, and which will affect the quality and quantity of natural capital, which in turn will impact on sustainability.

Approximately 55% of world CO_2 emissions increases arise from fossil fuel use, and the rest from deforestation, agriculture and other areas (IPCC SPM 1). Table 8.2 and figure 8.15 illustrate the structure and trends in CO_2 emissions (relating to consumption and flaring of fossil fuels), GDP and population.

In 2006, the top five countries, USA, China, Russia (all high consumers of coal) plus India and Japan accounted for nearly 55% of global CO_2 output from consumption and flaring of fossil fuels. World emission intensity (CO_2/GDP) at 2006 stood at 0.486Kg/$ of output, a 1% reduction on the figure for 2000, and a 44% reduction on the figure for 1980. World GDP, of course, had been marching onwards in that period, with a resultant significant increase in CO_2 production from fossil fuels, from 18,333 mn tonnes in 1980 to 28,003 mn tonnes in 2006. The world recession in 2008 likely has temporarily slowed this trend.

Figure 8.15 shows that, with the exception of France, which has significant nuclear input to electricity generation, most countries show only a small drop in carbon intensity. The levels for China and Japan have recently been rising, the former on a base of coal-fired power. Over the 20-year period to 2000 the world ratio has declined by less than 0.4% per annum. Policies to reduce this ratio include nuclear and renewable power, coupled with electric drive technology. While reduction in emissions ratios will assist in reducing man's impact on climate change, the other strategy is to reduce both per capita energy consumption and energy intensity/GDP, the trends of which are shown in figure 8.1. Here reduction has been more successful, with world energy intensity declining by about 2.4% per annum. Offset against this has been a rise in population of nearly 1.6% per annum.

Country	CO₂ Output mn toe pa	Carbon Intensity CO2/PEC ratio	Primary Energy Consumption mn toe pa	Energy Intensity Energy/GDP Kgoe/$	GDP PPP * bn $ pa	Population mn	GDP/Hd $ pa	PEC/Hd toe pa	Emission Intensity CO2/GDP Kg/$
USA	5696.8	2.4548	2320.7	0.2060	$11,265	299.8	$37,572	7.740	0.5057
China	5606.5	2.9840	1878.8	0.2163	$8,685	1311.8	$6,621	1.432	0.6455
Russia	1587.2	2.3472	676.2	0.4589	$1,474	142.5	$10,340	4.745	1.0771
India	1249.7	2.2087	565.8	0.1541	$3,671	1109.8	$3,308	0.510	0.3404
Japan	1212.7	2.2987	527.6	0.1491	$3,538	127.8	$27,694	4.129	0.3428
Germany	823.5	2.3625	348.6	0.1546	$2,255	82.4	$27,373	4.232	0.3652
Canada	538.8	1.9976	269.7	0.2652	$1,017	32.6	$31,178	8.269	0.5298
UK	536.5	2.3211	231.1	0.1322	$1,749	60.5	$28,888	3.818	0.3068
South Korea	476.1	2.1991	216.5	0.2135	$1,014	48.3	$20,992	4.482	0.4696
Italy	448.0	2.4327	184.2	0.1200	$1,535	58.9	$26,078	3.129	0.2919
France	377.5	1.3844	272.7	0.1609	$1,695	63.2	$26,819	4.314	0.2227
Australia	394.5	3.2208	122.5	0.1938	$632	20.7	$30,469	5.905	0.6242
Mexico	416.3	2.3461	177.4	0.1722	$1,030	104.8	$9,838	1.694	0.4039
Spain	327.7	2.2665	144.6	0.1382	$1,046	44.1	$23,731	3.280	0.3133
Brazil	332.4	1.4832	224.1	0.1518	$1,477	189.3	$7,800	1.184	0.2251
Rest of World	7978.9	2.2290	3579.5	0.2312	$15,482	2839.5	$5,452	1.261	0.5154
World 2006	28003.0	2.3853	11740.0	0.2039	$57,564	6536.0	$8,807	1.796	0.4865
World 2000	23850.0	2.5618	9310.0	0.1916	$48,600	6080.0	$7,993	1.531	0.4907

Table 8.2 CO₂ Output, Primary Energy Consumption, GDP and Population 2006

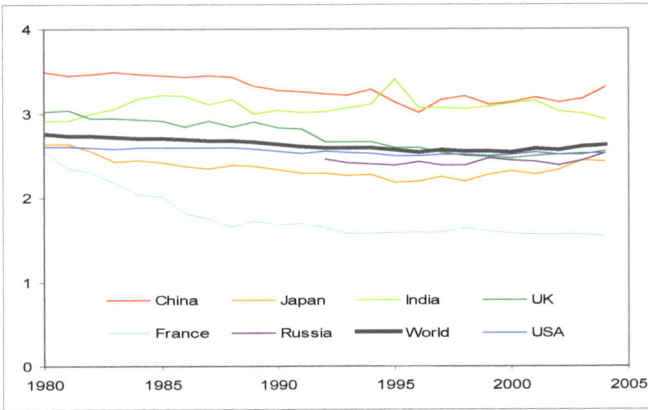

Figure 8.15 Carbon Intensity - ratio CO_2 emissions from consumption & flaring of fossil fuels / primary energy consumption (tonnes CO_2 / tonnes oil equivalent) 1980 – 2004. Sources: EIA, BP

The fourth assessment report of the IPCC sets out a range of scenarios of CO_2 emission reductions for increasing levels of global temperature and sea levels. Non of the scenarios includes for a rise in temperature of less than 2°C, and it is to be assumed that IPCC regard some rise in temperature as unavoidable. The strategy that achieves the lowest estimated increase in temperature and sea levels is the one that requires the greatest reduction in CO_2 emissions. It is inevitable that some countries may be more committed than others, and some may feel that their input should be less than others by virtue of history and their relative economic development. Table 8.3 abstracts data from the IPCC table:

Table 8.3 IPCC Stabilisation Scenarios

Stabilisation Scenarios	Change in Global CO_2 emissions in 2050 (percent of 2000 emissions)	Global Average Temperature Increase above pre-industrial at equilibrium	Global average sea level rise above pre-industrial at equilibrium
	%	°C	Metres
I	-85 to -50	2.0 – 2.4	0.4 – 1.4
II	-60 to -30	2.4 – 2.8	0.5 – 1.7
III	-30 to +5	2.8 – 3.2	0.6 – 1.9
IV	+10 to +60	3.2 – 4.0	0.6 – 2.4
V	+25 to +85	4.0 – 4.9	0.8 – 2.9
VI	+90 to +140	4.9 – 6.1	1.0 – 3.7

Source: IPCC AR4 Table 5.1

IPCC set out a number of areas of CO_2 mitigation, including energy (electric and non-electric) transportation, buildings, industry, agriculture, forest and waste management, along with cost estimates of strategies to meet potential savings (as a percent of GDP), which rise, the greater is the reduction in emissions to be achieved.

Temperature rises at scenarios III and upwards are likely unacceptable in terms of their impact on the environment.

It is intuitive to work back from a particular scenario to see the implications of a reduction in emissions. For example, in Scenario I a reduction of 85% in emissions against 2000 levels would imply (from table 8.2) world emissions in 2050 of 23,850 - 85% = 3,577.5 mn tonnes CO_2 (assuming that fossil emissions are in proportion to other emissions). Working back further, if no improvements in emissions intensity were achieved over 2000, this would imply, from the last column table 8.2, a level of world GDP of only 3,577/0.4907 = $7,290bn (compared to $48,600bn in 2000). Working back still further, if it is assumed that population grows to 9bn by 2050, then the level of world GDP per capita at 2050 would be $7,290/9 = $810 per capita, only 10% of the world level of $7,993 in 2000.

Clearly the above hypothetical example is somewhat unlikely; in particular, that reductions in the intensity ratios of both emissions and energy to GDP are taking place, as noted earlier. These may continue to improve, if actions are taken by member states of the global economy to invest in relevant capital stock and technologies to mitigate climate change. Rises in population size will, however, dilute significantly the level of GDP per capita.

Table 8.4 sets out a hypothetical example of progression from 1980 to 2050, assuming that world CO_2 output declined by 2.4% per annum from 2002 onwards to reach a level of 7,155 Gt pa, a reduction of 70%, not far from the IPCC maximum reduction of 85% set out at table 8.3. In passing it should be noted that table 8.2 shows that the carbon level rose 17.4% from 2000 to 2006. Thus already the world is behind on this projection. The other key assumptions made in table 8.4 are that carbon intensity reduces by 0.5% pa and energy intensity reduces by 2.0% pa, on a base of some continuation of efficiency improvements and some changes in habit. Last, it is assumed that world population continues to grow to reach a level of 9bn by 2050, in line with US Census Bureau projections. The calculations in the table are carried out in reverse, that is, from carbon target to primary energy consumption, then to GDP and last to GDP per capita.

Year	CO$_2$ Output (mn toe pa)	Carbon Intensity CO$_2$/PEC (ratio)	Primary Energy Consumption (mn toe pa)	Energy Intensity Energy/GDP (Kgoe/$)	GDP PPP (2000) (bn $ pa)	Population (mn)	GDP/Hd ($ pa)	PEC/Hd (toe pa)	Emission Intensity CO$_2$/GDP (Kg/$)
1980	18333.3	2.7602	6641.9	0.3122	$21,272.1	4451.0	$4,779	1.492	0.8618
1990	21426.1	2.6349	8131.6	0.2638	$30,820.3	5280.5	$5,837	1.540	0.6952
2000	23850.0	2.5618	9310.0	0.1916	$48,600.0	6080.0	$7,993	1.531	0.4907
2010	18700.0	2.4300	7695.0	0.1565	$49,170.0	6800.0	$7,231	1.132	0.3803
2020	14660.0	2.3100	6346.0	0.1279	$49,632.0	7500.0	$6,618	0.846	0.2954
2030	11490.0	2.1900	5247.0	0.1045	$50,222.0	8100.0	$6,200	0.648	0.2288
2040	9010.0	2.0900	4311.0	0.0853	$50,510.0	8600.0	$5,873	0.501	0.1784
205C	7155.0	1.9800	3614.0	0.0697	$51,823.0	9000.0	$5,758	0.402	0.1381

Table 8.4 Projections to 2050 of World Carbon Emissions*,
Primary Energy Consumption, GDP, Population and GDP per Capita
* From consumption and flaring of fossil fuels

It can be seen that world GDP per capita projected on this basis is likely to stagnate to 2050 if the world adheres to the emission targets recommended by IPCC. The actuality in between 2000 and 2050 is likely to be a good deal different to the projection above, taking into account 'same as usual' trends, delayed action to implement policies, non-adherence and the paths of individual countries. Some countries, in particular, Russia, China, Australia, Canada and USA have very high emissions intensity ratios (CO_2/GDP) to contend with. It is quite possible that reductions of emissions to low levels may not be achievable.

The overall emission intensity reduction from 2000 (the base level assumed by IPCC) to 2050 in the above example is from 0.4907 down to 0.1381, equivalent to a reduction of 71.8%. This figure is diluted in terms of GDP per capita, however, by an increase in world population of 48%. If population had remained level, GDP per capita would have risen.

A means of illustrating the impact on GDP per capita of varying emissions intensity across the range of emission change requirements stated by IPCC to 2050 is set out at figure 8.16.

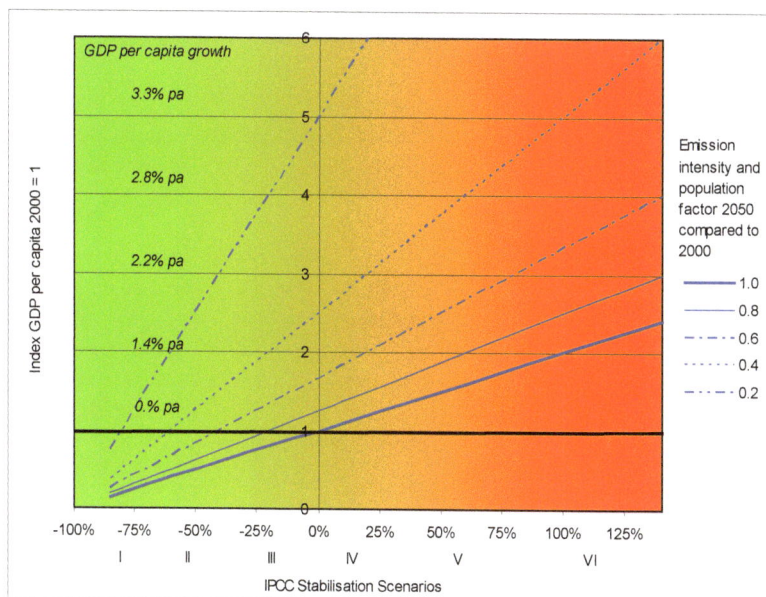

Figure 8.16 World GDP per capita as a function of Global Carbon Emissions Change

The range of IPCC stabilisation scenarios from table 8.3 is placed along the X-axis of the chart, with areas progressively shaded according to the increase in emissions and expected increases in global temperatures and sea levels. For low temperature rises only category I and perhaps II are viable, limiting policy consideration only to the left-hand part of the chart. An index of world GDP per capita is placed along the Y-axis, with the index of GDP per capita at year 2000 ($7,993 PPP constant prices) set at 1. While the author would no longer support projections of exponential growth, some approximate equivalents are set out for each index level to help readers. Thus, for example, an index of 5 in 2050, compared to 1 in 2000, would represent growth of 3.3% pa over 50 years between 2000 and 2050. As noted earlier, the position has deteriorated since 2000 with emissions at 2006 up by 17%, and population increased by 7.5%. Thus the world economy is currently marching towards the north-east part of the chart.

Splayed out across the chart is a range of world emission and population intensity factors, where 1 represents the position in year 2000. Table 8.4 shows this to be an emission intensity of 0.4907 combined with a population of 6.08bn. At the upper end a factor of 0.2, for example, might represent a combination of an increased population to 9bn combined with a reduction in emission intensity from 0.4907 in 2000 down to 0.0663 in 2050, the latter about half the level of 0.1381 given in the example at table 8.4.

It is a matter of evidence as to how far down emissions intensity can be driven. Clearly if all countries emulated France in its nuclear programme, undertook extensive electricity programmes in renewable sources, and switched road transport into electrically driven mode, then some advances beyond the thermodynamic limits of fossil-fired energy might be possible. Over a 40-year period changes to consumer habits might also be engineered, particularly if faced with unpalatable energy costs and loss of living standards. A drop in population projection below 9bn would also help. The alternative is for world humanity to settle for a high temperature rise. This entails very high global risks and a significant, irrevocable effect on the eco-system and life; and with the potential for a large reduction in levels of economic and social activity.

The above analysis adds weight to the conclusions concerning energy resources that the next few decades will likely follow a different path than one based on projecting GDP on an escalating, exponential basis. No account has been taken in this chapter of other factors, such as agriculture, fishing, water, and the move to an urban rather than a rural life. The latter factors will on the balance of probability likely add further weight to the analysis. There is likely to be a significant shift to the left in equation (4.3) of this book. Thus no single reaction kinetic, as set out at chapter 4, is likely to describe the forward path. The particular path that world economic development will follow is likely to be quite complex, given the interactions of the many parties, factors and systems involved, and that currently the world is still proceeding in the opposite direction to the recommendations of IPCC.

In 1972 a systems approach was attempted by Meadows (1972) for the Club of Rome, spawning a book *'The Limits to Growth'*, echoing perhaps concerns of Malthus [Malthus, T. R. (1798)]. *Limits to Growth* subsequently lost interest as economic development proceeded apace. It has recently been reviewed however by Turner [Turner, G. (2008)] who concluded that 30 years of historical data compared favourably with the "business-as-usual" "standard run" scenario of Limits to Growth, but did not compare favourably with other scenarios involving comprehensive use of technology or stabilizing behaviour and policies.

To summarise, climate change places a potential constraint on the forward path of output, in line with the thermo-economic thesis put forward in this book.

CHAPTER 9   THERMOECONOMICS AND SUSTAINABILITY

In neo-classical economics, sustainability in practice boils down to maximising utility in the form of consumption, usually of goods and services arising from economic production – so called economic capital. While natural capital can be included in this definition, unlimited substitution can be made between man-made and natural capital. Thus the only constraint placed upon consumers is that of a budget constraint – how much consumers have available to spend. This says little or nothing about outside systems – biological, ecological or other – which may have a bearing on consumers' choices. It is just assumed that these can if necessary be replaced or augmented over time by human ingenuity through the economic process.

If it is accepted that economic systems, and by definition human driven ones, seek to maximise entropy gain in some manner, subject to prevailing constraints, one might proceed further to consider the question of whether such a goal is consistent with sustainability of *both* economic and ecological systems.

This debate has been summarised by Brekke (1997) and Ayres (1998) as the difference between 'weak' and 'strong' sustainability, and the substitutability between natural capital and manufactured capital. Weak sustainability may be consistent with preserving the development of the human economic system, but this may be at the expense of reducing the effectiveness of the ecological environment, such as irrevocable resource depletion and ecosystem damage, which do not appear in traditional economic assessment. Strong sustainability on the other hand seeks to conserve natural capital as well as manufactured capital.

Economic management is still highly orientated towards giving humans what they want, a better life, usually enshrined as more GDP per capita, and certainly not less.

Significant variations also exist of individual per capita economic wealth, both at international and national levels. More than 60% of world GDP is attributed to countries counting for only 20% of world population (source: CIA 2006). It is a matter of evidence as to the geographic source of the resources contributing to this economic wealth. Japan for example has few natural resources, yet managed in the second half of the 20th century to build an economy based on processing resource value imported from elsewhere, such as metal ores and energy, combined with technology know-how.

Economic life is not equal, and it can be readily appreciated that those with a 'poorer' lot might aspire to catch up with those more fortunate, particularly if economic and technological tools are at hand to facilitate this. In the more recent past both China and India, with very large burgeoning populations, have grasped the economic genie that economics has offered other economies, and are setting about becoming the manufacturing hub of the world. This involves consuming increasingly large amounts of energy and other resources.

The evidence appears to indicate, however, that the world is gradually moving from a position where the only constraints to human economic growth have been those of human, monetary and economic capital, to a position where natural capital also is seen as being of prime importance as a constraint in the scheme of things. Moreover, because natural capital is complex and involves some long timescales, the current position has been developing for some time without humankind being generally aware of the situation. Climate change is an example where scientific evidence and accepted opinion indicates a constraint arising from man's interaction with the eco-system. There are other examples however, including resource depletion, fishing, farming, soil nutrient systems, deforestation, the hydrological cycle, the polar regions, the ozone layer, and waste accumulation and disposal. At a wider level humankind is only just beginning to comprehend how complex the eco-systems of planet Earth are, and what *Homo Economicus* may be doing to move them away from their equilibrium position, perhaps irrevocably. Numerous species of animate life are now threatened with decimation and some with extinction. Would Darwin, credited with the theory of evolution, have contemplated such a significant acceleration in extinction rates over such a short time?

Perhaps the biggest constraint that now exists is the sheer size of the human population that has grown on the back of economic capital based on non-renewable energy and renewable resources.

Neo-classical economics is not able to cater for this change in affairs, primarily because it is built on a circular flow of value between labour and economic capital, with little recognition that true value ultimately comes from resources and from natural capital, albeit that humankind provides the intelligence and management to harvest and garner these treasures. This has meant that until recently ecological economics has been regarded overwhelmingly as a side issue to mainstream economic thought. A considerable shift in accepted wisdom is now needed, away from a viewpoint that science has little or nothing to offer economics, towards structuring around disciplines incorporating terms and measures that relate to current accepted scientific views of the way the world works. A thermodynamic approach must be among the leading contenders for such a change in viewpoint.

From the thermo-economic development at chapter 3, it was shown that the notion of utility in an economic sense is closely related to that of entropy, and in chapter 4 it was also posited that maximising potential entropy gain is consistent with both natural processes and those arising in 'man-made' economic systems. A formal link between output flow and maximisation of potential entropy gain can only be established however if account is taken of constraints acting upon economic systems. These constraints take the form of factors that can affect flow of output; not just a budget constraint, but specific factors that can influence whether output goes up or down. Factors identified in this book include not only traditional economic ones such as money supply, production capacity and employment, but those involving natural capital, such as resource constraints – both renewable and non-renewable – and those constraints arising from the eco-system such as climate change.

From a thermodynamic viewpoint a process is more sustainable if less harmful losses of exergy/productive content occur. This is more likely for a renewable resource, as losses generated from consumption can gradually be replaced by those of new resources entering the flow, nurtured by nature and the sun. However the position with a renewable resource is not guaranteed, for if over time the rate of productive content/exergy usage inclusive of Second Law losses is greater than the rate of replacement of productive content/exergy, it is likely that the resource will gradually reduce in size, in the manner of a non-renewable resource. Strong sustainability on a thermodynamic basis therefore will only be achieved if humankind abstracts value that can be fully replaced by the sun through the interlinking eco-system and subsystems humankind relates to, such that future generations of humanity and life on earth can prosper in sustainable harmony, without the ecological system being compromised.

For example, draining marine life from some oceanic areas (including removing seabed life via trawling) to a level where the stock biomass has negative marginal growth, affects reproduction, with an ensuing decline in stock levels, which may also affect other higher and lower level species in the aquatic system. It follows that if the human population continues to grow alongside a declining marine stock, then at some point particular oceanic areas could become barren and not support aspects of human life.

The same logic follows regarding the use of land to grow food and nurture farm animals. Consuming even more energy to produce fertilisers may improve yields, but forever reducing arable land in favour of production of consumptive products is a negative a factor. No amount of exponential projection of human GDP will restore the source of food.

The net effect of a global policy that followed the weak sustainability path is that irretrievable natural resource depletion and ecosystem damage could escalate to a position where global constraints might force a severe retrenchment, if not something much more serious.

All of the above suggests that the key input to promoting strong sustainability may not be by humankind burning its way out by producing more GDP to feed its growing population in the hope that technology will solve the problem, but by consuming less. It is however quite contrary to human nature to reduce voluntarily an appetite for more, if not faced with an immediate constraint, and it is unlikely that individual nations and people would be willing to give up their way of living to the benefit of others unless, by international agreement and enforcement of such an agreement, each may be persuaded to reduce consumption if all suffer together. To date, international agreements on climate change policy, destruction of rainforests and harvesting of the oceans have not so far resulted in a significant change from the path of 'same as usual'.

Even if action were taken, it is human nature (the potential entropy maximisation principle again) that humankind having made a saving in one direction, would wish to go out and use the saving for something else. Thus just reducing expenditure on a non-renewable form of energy to replace it with another 'sustainable' form may leave GDP proceeding apace, but the threat of more population and short-term consumption of other resources such as food, and potential overloading of the eco-system would remain.

While global taxation might provide a means at the international level of persuading those who consume natural capital to pay and fund for their preservation, it is another matter to obtain acknowledgment of their liability and to enforce the economic redistribution involved. For example, it is a matter of evidence as to whether carbon taxes so far have reduced demand for fossil-fired energy, and attempts to enforce fishing quotas in parts of the world have been met by many trying to circumvent them. Further to the point, however, is the use that such taxes collected might be put. Redistributing the money to maximise human personal benefit might negate a goal to conserve the ecosystem.

In times past, the most effective occurrences that have persuaded both governments and populace to pull together to solve a problem have been in times of war and of recession/depression with reduced income. These were times when the populace experienced 'real' pain from a constraint or force, as opposed to being told that a 'threat' of pain might occur in some period ahead.

Human actions to ensure a high level of strong sustainability rest on the populace of all countries taking a much longer term view than hitherto has been the case, based on continued review and research into links between human and ecological systems. Nevertheless the advent of modern communications has meant that most people in the world are by now aware of the problem, even if singly and collectively they have so far done little to change their ways. Ingenuity and thought may play a part in the solution, but technology on it own will not provide an answer.

All this suggests that possibly a more likely outcome in the decades to come is eventually some reduced sustainability of natural capital, affecting carrying capacity, with an associated effect on human activity, a levelling out or even a decline in population and, over time, moderation of economic output to levels commensurate with prevailing constraints.

REFERENCES

Anderson, N. Sleath, J. (1999) 'New estimates of the UK real and nominal yield curves', *Bank of England Quarterly Bulletin*, November 1999.

Annila, A. Salthe, S. (2009) 'Economies evolve by energy dispersal', *Entropy*, 11, pp.606-633.

Ayres, R. U. van den Bergh, C. J. M. Gowdy, J. M. (1998) 'Viewpoint: Weak versus strong sustainability' 98-103/3, *Tinbergen Institute Discussion Papers.*

Ayres, R. U. Warr, B. Accounting for growth: The role of physical work'. *Centre for Management of Environmental Resources*, INSEAD.

Ayres, R.U. Martinas, K. 'Wealth accumulation and economic progress' *Evolutionary Economics* (1996) 6: pp. 347-359.

Ayres, R.U. Nair, I. (1984) 'Thermodynamics and economics'. Physics Today, November 1984.

Ayres, R.U. Warr, B. 'Two paradigms of production and growth' INSEAD.

Ayres, R.U. van den Bergh, J.C.J.M. Gowdy, J.M. (1998) Viewpoint: Weak versus strong sustainability. *Tinbergen Institute* Discussion Papers No 98-103/3.

Bacon, C. (2004) 'Practical portfolio performance measurement and Attribution', Wileys.

Baumgärtner, S. (2002) 'Thermodynamics and the economics of absolute scarcity' 2*nd World Congress of Environmental and Resource Economists*, June 24-27, 2002, Monterey, CA, USA.

Baumgärtner, S. (2004) 'Temporal and thermodynamic irreversibility in production theory' Economic Theory, 26, pp. 725-728

Baumgärtner, S. Faber, M. Proops, J.L.R. (1996) 'The use of the entropy concept in ecological economics', Faber et al. (1996: Chap. 7).

Brekke, K. A. (1997) 'Economic growth and the environment: On the measurement of income and welfare.' *Edward Elgar*, Cheltenham.

Bryant, J. (1979) 'An equilibrium theory of economics', *Energy Economics*, Vol 1, No 2, pp. 102-111.

Bryant, J. (1982) 'A thermodynamic approach to economics', *Energy Economics*, Vol.4, No. 1, pp. 36-50.

Bryant, J. (1985) 'Economics, equilibrium and thermodynamics', Workshop Energy & Time in the Economic and Physical Sciences, Elsevier Science Publishers B.V. pp. 197-217.

Bryant, J. (2007) 'A thermodynamic theory of economics', *International Journal of Exergy*, Vol.4, No. 3, pp. 302-337.

Bryant, J. (2008) 'A thermodynamic approach to monetary economics', *Vocat International Ltd*, www.vocat.co.uk.

Bryant, J. (2008) 'Thermodynamics and the economic process. An application to World energy resources and climate change', *Vocat International Ltd,* www.vocat.co.uk.

Candeal, J.C. De Migule, J.R. Indurain, E. Mehta, G.B. (1999). 'Representations of ordered semigroups and the physical concept of entropy'. Preprint (1999).

Candeal, J.C. De Migule, J.R. Indurain, E. Mehta, G.B. (2001) 'On a theorem of Cooper'. *Journal Mathematical Analysis & Applications. Math.* Anal. Appl. 258 (2001) pp. 701-710.

Candeal, J.C. De Migule, J.R. Indurain, E. Mehta, G.B. (2001) 'Utility and Entropy' *Economic Theory*, Springer-Verlag, 17, pp. 233-238.

Chakrabarti, B. K. Chatterjee, A. 'Ideal gas-like distributions in economics:Effects of savings propensity'. arXiv:cond-mat/0302147

Chakraborti, A. Patriarca, M. (2008) 'Gamma-distribution and wealth inequality' International Workshop on: Statistical Physics Approaches to Multi-disciplinary Problems, IIT Guwahati.

Chen, J. (2007) 'Ecological Economics: An analytical thermodynamic theory' University of Northern British Columbia, Canada

Chen, J. (2007) 'The physical foundations of economics' www.worldscientific.com

Cleveland, C. J. (2008) 'Biophysical economics' www.eoearth.org.

Cobb, C.W. Douglas, P.H. (1928) 'A theory of production', *American Economic Review*, 18 (supplement), pp. 139-165.

Costanza, R, Hannon, B. (1989). 'Dealing with the mixed units problem in ecosystem network analysis'. In F Wulff, JG Field, KH Mann (eds) *Network Analysis in Marine Ecology: Methods and Applications*, pp. 90-115, Berlin: Springer.

Costanza, R. (1980) 'Embodied energy and economic valuation' *Science* Vol 210. no. 4475. pp. 1219-1224.

Daly, H. E. (1991) 'Steady state economics: Second edition with new essays' *Island Press*, Washingron, DC.

Daly, H. E. (1992) 'Is the entropy law relevant to the economics of natural resources? Yes, of course it is!', Journal of Environmental Economics and Management, 23, pp. 91-95.

Daly, H. E. (2005) 'Economics in a full world'. *Scientific American*, Vol 293, Issue 3.

Debreu, G. (1954). 'Representation of a preference ordering by a numerical function'. Decision Processes, New York: Wiley pp. 159-165.

Debreu, G. (1959) 'The theory of value: An axiomatic analysis of economic equilibrium'

Dragulescu, A. Yakovenko, V. M.(2000) 'Statistical mechanics of money' *Eur. Phys.J. B 17, pp. 723-729*

217

Dragulescu, A. Yakovenko, V. M.(2001) 'Exponential and power law
    probability distributions of wealth and income in the United Kingdom
    and the United States'. Physica A, 299 (2001) pp. 213-221.
Estrella, A. Trubin, R.T. (2006) 'The Yield Curve as a Leading Indicator:
    Practical Issues', *Federal Reserve Bank of New York*, Vol 12, No 5.
Eyring, H. (1935). 'The Activated Complex in chemical Reactions'. *J.
    Chem. Phys.* 3: 107. doi:10.1063/1/1749604.
Eyring, H. Polanyi, M. (1931). *Z. Phys. Chem. Abt* B 12:279.
Farmer, J.D.. Shubik, M. Smith, E. (2008) 'Economics: the next physical
    science?' Santa Fe Institute
Ferrero, J. C. (2004) 'The statistical distribution of money and the rate of
    money transference' Physica A: Vol 341, pp. 575-585.
Fisher, I. (1892) 'Mathematical investigations of the theory of values and
    prices'. Transactions of the Connecticut Academy of Arts and Sciences
    9, pp. 11-126.
Frederick, S. Loewenstein, G. O'Donoghue, T. (2002) 'Time discounting
    and time preference: A critical review' Journal of Economic Literature,
    Vol 40, No. 2, (June) pp.351-401.
Friedman, M. (1956), 'The Quantity Theory of Money: A Restatement' in
    Studies in the Quantity Theory of Money, edited by M. Friedman.
    Reprinted in *The Optimum Quantity of Money* (2005), pp. 51-67.
Georgescu-Roegen, N. (1971) 'The entropy law and the economic process',
    Harvard University Press, Cambridge MA.
Georgescu-Roegen, N. (1979) 'Energy analysis and economic valuation",
    Southern Economic Journal, 45, pp. 1023-1058.
Goldberg, R.N. Tewari, Y.B. Bhat, T.N. (1993-1999) 'Thermodynamics of
    enzyme-catalysed reaction Parts 1-5' *Journal of Physical and Chemical
    Reference Data.*
Green, J. Baron, J. (2001) 'Intuitions about declining marginal utility'
    Journal of Behavioural Decision Making. 14, pp. 243-255.
Grubbeström, R. W. (1985) 'Towards a generalised exergy concept'. Energy
    & Time in the Economic and Physical Sciences, North Holland,
    Amsterdamm, pp. 225-242.
Hannon, B. (1973) 'An energy standard of value', *Annals of the American
    Academy,* 410: pp. 139-153.
Hansen, A (1953) A Guide to Keynes, *New York: McGraw Hill.*
Harrison, S. et al. (2007) 'Interpreting movements in broad money', *Bank of
    England Quarterly Bulletin,* 2007 Q3.
Hau, J. L. Bakshi, B. R. (2004) 'Promise and problems of emergy analysis',
    Ecological Modelling, Ohio State University, Columbus, Ohio, 178: pp.
    212-225.
Hicks, J. (1932) 'The theory of wages'. *Macmillan,* London.

218

Hubbert, M.K. (1956) 'Nuclear energy and the fossil fuels'. *American Petroleum Institute*, Spring meeting, Southern District.

Kamps, C. (2004) 'New estimates of government net capital stocks for 22 OECD countries 1960-2001'. *Kiel Institute for World Economics.*

Kapolyi, L. 'Systems analysis of bio-economy entropy and negentropy in biopolitics' *Systems International Foundation*, Hungary.

Kay, J. Schneider, E. (1992) ' Thermodynamics and measures of ecosystem integrity', *Ecological indicators*, Elsevier, New York, pp. 159-191.

Keynes, J.M. (1936), 'The General theory of employment, interest and money' London: Macmillan (reprinted 2007).

Klein, L. (1962) 'An introduction to econometrics'. Prentice-Hall, Inc. pp. 154-156.

Koopmans, T. C. Ed (1951) 'Activity analysis of production and allocation'. New York, John Wiley & Sons.

Laherrere, J. (2005) 'Forecasting production from discovery'. *ASPO.* Lisbon May 19-20, 2005.

Laherrere, J. (2006) 'Fossil fuels: What future?' *China Institute of International Studies.* Workshop October 2006, Beijing.

Leontief, W. W. (1986) *'Input-output economics'.* 2nd ed., New York: Oxford University Press, 1986.

Lezon, T.R, Banavar, J. R. Cieplak, M. Fedoroff, N. (2006) 'Using the principle of entropy maximization to infer genetic interaction networks from gene expression patterns'. PNAS vol 103, no.50 pp. 19033-19038.

Lisman, J. H. C. (1949) 'Econometrics statistics and thermodynamics', *Netherlands Postal and Telecommunications Services*, The Hague, Holland, Ch IV.

Mackay, D.J.C. (2008) 'Sustainable Energy – without the hot air', *UIT Cambridge Ltd,* ISBN 978-0-9544529-3-3.

Malthus, T. R. (1798) 'An essay on the principle of population'.

Mahulikar, S. P. Herwig, H. (2004) 'Conceptual investigation of the entropy principle for identification of directives for creation, existence and total destruction of order' *Physica Scripta*, vol 70, pp. 212-221.

Martinás, K. (2000) 'An irreversible economic approach to the theory of production' *Systems & Information Dynamics*. Vol. 7, Issue 4.

Martinás, K. (2002) 'Is the utility maximization principle necessary? *Post-autistic economics review*, issue no. 12, March 2002.

Martinás, K. (2005) 'Energy in physics and in economy' *Interdisciplinary Description of Complex Systems* 3(2), pp. 44-58.

Martinás, K. (2007) 'Non-equilibrium economics' *Interdisciplinary Description of Complex Systems* 4(2), pp. 63-79

Meadows, D.H. Meadows, D. L. Randers, J. Behrens III, W.W. (1972) 'The limits to growth'. *Universe Books.*

Mirowski, P. (1989) 'More heat than light: Economics as social physics' *New York: Cambridge University Press.*

Mirowski, P. (2002) 'Machine dreams: Economics becomes a cyborg science' *New York: Cambridge University Press.*

Nelson, C.R., Siegel, A.F. (1987) 'Parsimonious modelling of yield curves' *Journal of Business,* 60(4), pp.473-489.

Neumann, J. von. Morgenstern, O. 'Theory of Games and Economic Behavior'. Princeton, NJ. Princeton University Press. 1944 sec.ed. 1947.

Odum, H. T. (1971) 'Environment, power & Society', *Wiley,* London.

Odum, H. T. (1973) 'Energy, ecology and economics' Ambio 2, pp. 220-227.

Odum, H. T. (1998) 'Emergy evaluation', *International Workshop on Advance in Energy Studies*, Porto Venere, Italy.

Ozawa, H. Ohmura, A. Lorenz, R. D. Pujol, T. (2003) 'The second law of thermodynamics and the global climate system: A review of the maximum entropy production principle'. American Geophysical Union, Reviews of Geophysics, 41, 4 / 1018

Patterson, M. (1998) 'Commensuration and theories of value in ecological economics' *Ecological Economics,* 25, pp. 105-125.

Phillips, A. W. (1958) 'The relationship between unemployment and the rate of change of money wages in the United Kingdom 1861-1957' Economica 25 (100) pp. 283-299.

Pikler, A. G. (1954) 'Optimum allocation in econometrics and physics', *Welwirtschaftliches Archiv.*

Proops, J. R. (1985) 'Thermodynamics and economics: from analogy to physical functioning'. Energy & Time in the Economic and Physical Sciences, North Holland, Amsterdam, pp. 225-242.

Purica, I. (2004) 'Cities: Reactors of economic transactions'. Romanian Journal of Economic Forecasting. Vol 1. Issue 2 pp. 20-37.

Raine, A. Foster, J. Potts, J. (2006) The new entropy law and the economic process', *Ecological Complexity*, 3, pp. 354-360.

Ruth, M. (1993) 'Integrating economy, ecology and thermodynamics'. Kluwer Academic, Dordecht.

Ruth, M. (2007) 'Entropy, economics, and policy' Universitat Bremen, artec-paper Nr. 140

Samuelson, Paul. (1937) "A Note on Measurement of Utility," *Rev. Econ. Stud.* 4, pp. 155-61.

Samuelson, P. A. (1947) 'Foundations of economic analysis', *Harvard University Press.*

Samuelson, P. A. (1964) 'An extension of the Le Chatelier principle' *Econometrica*, 28, pp. 368-379.

Samuelson, P. A. (1970) 'Maximum principles in analytical economics', *Nobel Memorial Lecture*, Massachusetts Institute of Technology, Cambridge, MA.

Samuelson, P. A. (1980) 'Economics' 11th edition, McGraw-Hill.

Schneider, E. D. (1987) 'Schrödinger shortchanged', *Nature* 328, 300.

Schneider, E. D. Kay, J. J. (1995) 'Order from disorder: The thermodynamics of complexity in biology', *What is life: The next fifty years. Reflections on the future of biology*, Cambridge University Press, pp. 161-172.

Schrödinger, E. (1944) 'What is life?' London: *Cambridge University Press.*

Shannon, C. E. Weaver, W. (1949) 'The mathematical theory of information', University of Illinois, P, Urnaban, Ill. (1949).

Smith, E. Foley, D. K. (2002) 'Is utility theory so different from thermodynamics?', *Santa Fe Institute.*

Smith, E. Foley, D. K. (2004) 'Classical thermodynamics and general equilibrium theory', *Santa Fe Institute.*

Soddy, F. (1934) 'The role of money', *Routledge, London.*

Söllner, F. (1997) 'A re-examination of the role of thermodynamics for environmental economics', *Ecological Economics*, 22. pp. 175-201.

Solow, R. M. (1956) A contribution to the theory of economic growth'. *Quarterly Journal of Economics*, 70, pp. 65-94.

Solow, R. M. (1957) 'Technical change and the aggregate production function'. *Review of Economics & Statistics.* 39. pp. 312-320.

Solow, R. M. (1974) 'Intergenerational equity and exhaustible resources. *Review of Economic Studies,* vol 41: pp. 29-45.

Sousa, T. Domingos, T. (2005) 'Is neoclassical microeconomics formally valid? An approach based on an analogy with equilibrium thermodynamics', *Ecological Economics,* in press.

Sousa, T. Domingos, T. (2006) 'Equilibrium econophysics: A unified formalism for neoclassical economics and equilibrium thermodynamics' P*hysica A*, 371 (2006) pp. 492-512.

Swenson, R. (1988) 'Emergence and the principle of maximum entropy production'. Proceedings of 32nd Annual meeting of the International Society for General Systems Research, 32.

Svensson, L. (1994) 'Estimating and interpreting forward interest rates: Sweden 1992-94', *IMF Working Paper,* No 114.

Swenson, R. (2000) 'Spontaneous order, autocatakinetic closure, and the development of space time'. Annals New York Academy of Science vol 901, pp. 311-319.

Taylor, J. B. (1993) 'Discretion versus policy rule in practice', *Carnegie-Rochester*, Conference series on public policy 39, December, pp. 195-214.

Turner, G. (2008) 'A comparison of the limits to growth with thirty years of reality'. *Global Environmental Change,* vol 18, Issue 3, pp. 397-411.

Udgaonkar, J.B. (2001) 'Entropy in Biology' Resonance September 2001.

Waggoner, D. (1997) 'Spline methods for extracting interest rate curves from coupon bond prices', *Federal Reserve Bank Atlanta,* Working Paper No 97-10.

Wall, G. (1986) 'Exergy - A useful concept' *Chalmers Biblioteks Tryckeri,* Goteborg 1986.

Warr, B. Ayres, R.U. (2002) 'An introduction to REXS a simple systems dynamics model of long-run endogenous technological progress, resource consumption and economic growth'. INSEAD.

Warr, B. Schandl, H. Ayres, R.U. (2007) 'Long term trends in resource exergy consumption and useful work supplies in the UK 1900 - 2000' CSIRO Sustainable Ecosystems.

Yuqing, H. (2006) 'Income Distribution: Boltzmann analysis and its extension' Physica A: Statistics mechanics and its applications. Vol 377, Issue 1. pp. 230-240.

Zittel, W. Schindler, J. (2007) 'Coal: Resources and future production'. *Energy Watch Group,* EWG-Series No1/2007.

DATA SOURCES

BP Statistical Reviews
Energy Information Administration (EIA), US Department of Energy
IEA
Intergovernmental Panel on Climate Change:
    Third & Fourth Assessment Reports.
OECD
Penn World
United Nations
USA Census Bureau
www.bea.gov
www.statistics.gov.uk
www.federalreserve.gov

LIST OF SYMBOLS

Symbol	Thermo-Economic	Thermodynamic
t	Time	Time
P	Price	Pressure
V	Volume flow rate	Volume (3-D)
N	Number of stock units	Number of molecules
v (=V/N)	Specific Volume Rate	Specific Volume (1/density)
G (=PV)	Value flow rate	Energy
k	Productive Content/unit	Boltzmann constant
Nk	Stock Productive Content	n.a.
T	Index of Trading Value	Temperature
S	Entropy	Entropy
s (=S/N)	Entropy per unit	Entropy per unit
F	Free Value (flow)	Free Energy (Helmholtz)
X	Free Value (flow)	Free Energy (Gibb)
f (=F/N)	Free Value per unit	Free Energy per unit
C_v	Specific Value (Const volume)	Specific Heat (Const volume)
C_P	Specific Value (Const price)	Specific Heat (Const pressure)
n	Elastic Index	Index Expansion/Compression
γ (=C_P/C_V)	Isentropic Index	Isentropic Index
Q	Entropic Value flow	Heat Supplied/lost
W	Work Value flow	Work Done
U	Internal Value (stock value)	Internal Energy
u (=U/N)	Internal Value per unit	Internal Energy per unit
ψ	Equilibrium Constant	Equilibrium Constant
ξ	Lifetime Ratio	n.a.
ω	Value Capacity Coefficient C_v/k	C_v/k

INDEX

224

www.ingramcontent.com/pod-product-compliance
Lightning Source LLC
Chambersburg PA
CBHW042146220326
41599CB00003BB/4